AF822281

Number and the Nature of Mathematics

Joseph Bergin, Ph. D.

Published by Slant Flying Press, 2023

ISBN: 978-1-940113-07-4

Audience: This work is intended for anyone interested in pursuing mathematical studies. I found the material especially valuable as a first year undergraduate. Teachers of mathematics should also find it useful to give themselves background and basic understanding to the things that they teach students – even very young students. No background is assumed beyond basic arithmetic.

Dedication: This book is dedicated to the memory of Clarence J Wallen, S.J., who set me on my mathematical journey.

Foreword: In the Spring of 1962, I was a first year student at Loyola University of Los Angeles, studying math. I was fairly new to math (and to study, for that matter), but Father Wallen, as we knew him, took a few of us under his wing in a special course, taught with his own notes. He took us through the development of the number system, from the Natural Numbers to the Complex Numbers, one step at a time. Along the way we followed a generally historical path, seeing the introduction of new ideas and becoming familiar with axiom systems and their implications.

This course gave me the best idea of the nature of mathematics that I would gain until late in graduate school. Three of us took the course, and I believe that we were the first to do so. All of us, myself, Alexander Hahn, and Michael Cullen, all went on to earn doctorates and to teach the rest of our careers. Mike went back to Loyola to teach Math and become department head. He died several years ago. Alex was head of the Mathematics Honors Program at Notre Dame where he has spent most of his adult life. I switched to Computer Science and taught that for most of my career, never losing the abstract view that I learned from *Clancy,* as his friends knew him (though we never dared to utter it).

Acknowledgements:

Graphs: https://www.desmos.com/calculator. Chinese Abacus, ca 1925: Smithsonian Institution negative number DOR 2010-0104. Maya Calendar: Matthew Bisanz at the Smithsonian. Sheep Sculpture Photo: Wouter Hagens. Quipu: Meyers Konversationslexikon of 1888 - Public Domain. Old Scroll: Arvin61r58, OpenClipart.org. Archer: GDJ, OpenClipart.org.

Ian Benson, Director, Sociality Mathematics CIC made valuable suggestion to the text and direction. Lisa Bergin gave me advice on some of the ideas here. Cover photography by the author.

About the Author

Joseph Bergin earned his Ph. D. in Mathematics in 1972 from Michigan State University. Since then he has been an educator, first in mathematics, but mainly in computer science. The job market for mathematicians in 1972 was grim. and those who wanted to teach needed to be inventive about building a career. Computer science turned out to be a good fit and while he never lost his view of himself as a mathematician, he focused mostly in helping new students become proficient in computing.

As a youngster, Joe was an indifferent student. He remembers a parent-teacher conference in which the school suggested holding him back a year as he had not yet memorized his multiplication tables. He recalls that he finally mastered them shortly after earning his doctorate. School was not the focus of his life then, and no subject really caught his attention, though he asked a lot of questions of everyone from a young age. He wasn't dumb, but mostly bored by school. It wasn't until a geometry course in secondary school that he found anything interesting to study.

Though he didn't really recognize it at the time, the course described in the dedication, and the subject of this book, was his first introduction to the nature of mathematics. But it took longer still, his fifth year of doctoral study, before he really grasped the true nature of mathematics. That seems to be crazy. It shouldn't have been like that. He had studied math intensely for all of his adult life without really knowing what it was *about*. What is it that mathematicians *do* when they do *Mathematics*.

Contents

About the Author....iv
Introduction....viii
1 Mathematics: Misconceptions....1
2 Early Counting – Speculation....3
First Steps....3
Then, The Babylonians, Egyptians, and The Greeks....5
And Elsewhere....5
3 Natural Numbers and Axiom Systems....7
Axioms....7
Definitions and Properties....8
Addition....19
Properties of Addition - Proof Based on Axioms....23
The Associative Law....24
The Law of Substitution....24
The Commutative Law....25
What is Mathematics: A First Look....27
The Law of Cancellation....28
Summary So Far....29
4 An Introduction To Sets And A Bit Of Logic....31
Sets - Definitions and Properties....31

Simple Logic.......37
A Note About Infinity.......41
5 Zero. A First Extension – The Whole Numbers.......43
Definitions and Properties.......43
A new relationship: Less-Equals.......46
Some Terminology.......49
Multiplication.......52
What is Easy, What is Hard, and Mathematical Insight.......56
Algebra, Variables, Unknowns, and How Mathematics is Not Computer Science.......58
Some Interesting Sums.......59
Representing Whole Numbers.......60
Summary So Far.......63
6 The Integers.......65
Equivalence Relations and Equivalence Classes.......67
Definitions and Properties.......68
Generalization and the Nature of Mathematics - Foreshadowing Abstract Algebra.......76
Applied Mathematics: A Simple Example.......78
Subtraction.......81
A Note on Proof.......82
Substitution and Cancellation Laws in Z.......83
Integer Division and Remainders.......85
A Note on Mathematical Insight.......91
Summary So Far.......94
7 The Rational Numbers.......97
Definitions and Properties.......98
Denseness and Distance in the Rational Numbers.......105

Cardinality of the Rational Numbers 107
Decimals and the Rational Numbers 109
Summary So Far 110
8 Infinite Sequences 113
Definition and Properties 113
Infinite Series 120
9 The Real Numbers 123
Cauchy Sequences 124
Definition and Properties 126
Cardinality of the Real Numbers 137
Foreshadowing: Topology 139
Foreshadowing – Calculus and Analysis 140
Summary So Far 149
10 Polynomials 151
Definition and Properties 151
11 The Complex Numbers 155
Definition and Properties 155
Foreshadowing – Trigonometry 158
Foreshadowing – Matrix Theory 161
Summary So Far 166
12 Mathematics 167
What Do Mathematicians Do? 167
A Simple Example 169
13 Ideas for Studying Mathematics 181
14 Further Reading 185

Introduction

While the concept of number doesn't completely cover the idea of mathematics, it has been an important component. Nearly everyone today learns something about numbers and a bit about math itself, but few people truly understand the fundamental idea of mathematics. So, what is Mathematics? Really? Here we hope to shed some light on that question.

We, you and I, will use the development of number systems to try to get a handle on what it is that mathematicians do. They don't memorize facts like you did at age 10, or so. They don't *just* prove theorems as you may have done in secondary school or university. It is actually quite different. Before one can prove a theorem they need to discover what the theorem actually is. It may not be involved with number at all. But mathematicians do prove theorems: statements that can be derived logically from simpler known truths. And here you will have an introduction to theorem proving if you haven't seen it previously. But that isn't the *essence* of math.

Some of what mathematicians do is deadly boring, but necessary. Some of it is wildly exciting. The boring stuff comes when you need to verify some rules and prove theorems using already-understood techniques. The exciting stuff happens when you find a question that no one has answered sufficiently and begin to explore it. It is a search for mathematical truth that may begin in a very dark place. None of your existing tools shed any light on the topic you'd like to explore. Often frustrating. Occasionally exhilarating. Historically it has even been controversial, inviting ridicule from prominent mathematicians of the day, and only later becoming generally accepted.

The development we will use here is primarily a product of 19th and 20th century thought. Prior to that, the tools used by mathematicians, logicians, and philosophers were more informal. We will trace the development from the counting numbers, the Natural Numbers, through the Complex Numbers. The latter are needed to do the engineering used in your smartphone. We will use a sequence of extensions starting with an axiomatic base. The axioms are supplemented by definitions of interesting things such as operations, like *addition,* and relations, such as *less than*. We will stick pretty close to this path, but will occasionally point to additional topics that depend on or are related to the ideas of the main thread.

Some of it will be a bit tedious, since many of the rules we use are proven with similar techniques. Once you have proven six theorems and see that there are six more to go with the same sort of calculations or arguments, it gets a bit boring. Mistakes are often made in such situations. But then something new comes along to refresh the game. And don't get the idea that mathematics is finished in any sense. It is an open and expanding world of potential study.

Mathematics is central to a classical education. In the Medieval university there were seven basic subjects, all taught in Latin: Arithmetic, Geometry, Astronomy, Music Theory, Grammar, Logic, and Rhetoric. Liberal Arts colleges such as Loyola try to give students a grounding in most of these as well as more modern ideas. I studied all but Astronomy as an undergraduate, though not all to the same depth. More advanced scholars in the university of year 1200 or 1300 would study Moral Philosophy, Metaphysics, and Physics. I too studied these in the 20th century. But it was the rigorous way of thinking provided by the basic subjects, especially maths, that enabled the later studies.

This book has ideas drawn largely from arithmetic and logic, but will provide excursions to other areas of math. It does not, however, give a complete treatment of any single topic. We will build on ideas established early to develop later topics and so will focus on essentials needed for that development. Along the way we will leave much for you to explore. Think of it as a long hallway with many, many, rooms off to the side. We will spend time in some of these rooms, but will only glimpse a few of the others. Some of those rooms lead to other hallways, each with additional rooms. You will be able to explore those elsewhere.

I should note at the start that the ideas presented here about the nature of mathematics are personal. Others may disagree. Perhaps my view is too narrow. Perhaps it is too specialized. But if it gives you a small window into this vast world then it will be useful for both of us. I hope you enjoy the view.

Occasionally we will mark a phrase with the symbol ✚. This indicates that a Google or Wikipedia search will yield additional information, though that is often true of many things we don't mark.

1 Mathematics: Misconceptions

What is mathematics? Everyone studies it but few people really know the answer to this question. I was in my fifth year of graduate study in mathematics before I really understood and I already had two degrees in Mathematics, a B.A. and an M.A. It was, really, about time I understood. It took still longer to really *become* a mathematician.

What most people study, and what they think of as mathematics, are just facts. These are the *leavings* of mathematicians. The old stuff that our predecessors once thought interesting and important. Even the processes, algorithms, that you learned, how to multiply large numbers for example, are just a more sophisticated kind of fact. The facts are often useful to mathematicians and non-mathematicians alike, but they are not mathematics. Perhaps they once were, but no more.

Those who have advanced a bit, say a typical undergraduate studying math or science and now taking Calculus, may think of mathematics as theorem proving: establishing the facts. But that isn't it, really, either. You need to climb quite a bit higher before you reach real mathematics. Perhaps you find theorem proving tedious and even boring. Mathematicians often feel the same. I am now not certain that I was really a mathematician even on the day I earned my doctorate, though I became one soon after.

If you actually *like* mathematics, then you may be very excited to learn something new, say, quaternions(+). Mathematicians often feel the same, but that isn't the essence either. If it is something really new, just discovered, then mathematicians are likely to be happier still, but mostly as it may reveal a new tool for their own work.

So, like the early passages of the fourth movement in Beethoven's Ninth Symphony(+), I've tried to show you what mathematics is *not*. But you will need to wait a bit to see what I think mathematics really *is*. Sorry, no easy answers here. But at least you won't have to actually become a mathematician in order to understand, to a degree, what mathematics is.

But, like any craftsman or craftswoman, a mathematician needs good tools. You have possibly gotten some experience with these tools. Abstraction is a good, solid, hammer. Logic a good, and essential, cutting tool. Facts are needed so that you have something to stand on while you work: scaffolding. Cleverness also helps, as does experience. Theorem proving helps to strengthen your mental muscles. Oddly enough, an aesthetic sense and a love of elegance is truly helpful. But these tools are also useful for other things. Computer scientists need them all, of course, and they may need mathematics, or not, depending on their particular focus. But none of these tools explain the essence of it.

My own background is, perhaps, informative. I was always a curious person as a youngster, always asking questions. But I was an indifferent student. I came late to math and even to an interest in school at all. The first course that ever actually meant something to me was Geometry in secondary school. I loved trying to solve the various exercises relating to triangles and other simple geometric figures. Up until then my grades were not very good, but this course helped cause a turn around in my attitude. I finished secondary school rated highly by my teachers and did well thereafter at University and graduate school, though I stumbled a bit. My background was so poor, in fact, that I finally taught myself the multiplication tables after earning the doctorate. It was interest and hard work, along with good teaching, that I credit for my success, rather than any innate ability. Just like my Geometry experience, I developed deep insight into the behavior of functions by working through a huge number of examples, graphing them by hand, so that I could deduce the behavior of many functions just by looking at a formula.

My fellow students in graduate school thought me to be smart. But that was only because I asked a lot of questions in class. I've learned that if you want to *seem* to be smart, ask a lot of questions. If you want to actually *be* smart, listen to the answers.

2 Early Counting – Speculation

First Steps

Most of you have probably heard somewhere that there exist some primitive people whose only counting concepts are *one, two, many*. Likely this was a common occurrence with primitive humans. If you were in a hunting society then you had to approach the prey. The difference between one and two buffalo is probably significant in your hunting tactics, but once there are three or more the technique might not change much. Likewise if you were the prey instead, the difference between one wolf and two may make a difference, but once there were more, the differences might not be significant. Certainly the difference between 200 and 201 buffalo (or wolves) is probably beyond caring.

If you have the concept of *many*, then "many many" is probably understandable to your hunting band, without the need to be more precise. Even being able to put a precise value to the number of people in your group is not very useful, assuming the number is small enough that you can "enumerate" them by name. So, I speculate, the development of number in primitive humans followed the need, and the initial need was slight. One possible exception to that idea is counting days in a moon cycle or in a season. Even hunter-gatherers need to be aware of the seasons when foods become available. Early agriculturalists need awareness of seasons even more. This eventually let to sophisticated calendar systems, that we won't explore here.

However, once we reach the stage of larger groups, kings, kingdoms, armies, taxes, and all the rest, humans had a need to count. How many people owe us taxes? How many sons can the miller contribute to the army? Things like that. The ancient Egyptians and the ancient Inca had very sophisticated counting systems for things like this. By the time of the Greek philosophers, certainly, it was recognized that the "Natural Numbers" are infinite in extent.

Zeno's Paradox (ca. 450 BCE) indicates a highly sophisticated view of number. Motion is impossible, he said. If you shoot an arrow at a wall, first it needs to go half way. Then again, it needs to go halfway. After that, halfway again. Infinitely. It has no end. Therefore it can't reach the wall, and therefore motion is impossible. A bit later (ca. 300 BCE), Euclid had very well developed ideas about number as well as geometry.

Oddly, it seems, zero came to be recognized as a *number* only late in the game. The Egyptians had a symbol for zero, but not as a number. Rather it was a concept in their accounting practice. It was a way of expressing "nothing", but not as something that "came before 1" in a counting scheme. That realization came to India and independently to the Olmec peoples in Mesoamerica when they needed it to develop a positional representational system. But that was much later.

By the 18th century, Europeans had sophisticated ideas about number, and it was at about this time that mathematics started to become formalized. The reason for formalizing had a lot to do with trying to resolve paradoxes. It was also a new realization that the facts of arithmetic, didn't just "exist" but that some could be derived from others using simple logical tools. This led to a desire to reduce much of mathematics to the simplest formulations and then attempt to derive the rest from that.

In this book we will follow along some of this development, taking many excursions along the way. The reader should gain some appreciation about how mathematicians think and why they develop new mathematics, which is the essence of math. The *facts* are of relatively small importance., though they feed our enquiring minds.

However, the earliest roots of mathematics weren't all about counting. Shape was another important component. It led to geometry, which was already highly developed in ancient Greece and elsewhere. The relationship between number and shape was, at one time, an important insight. When geometers discovered that the diagonal of a square with side 1 couldn't be any Rational Number it caused great concern and that, eventually, led to new mathematics. Somewhat at the intersection of counting and geometry is the notion of measuring. Something is big. How big? There is a distinction between *how many* and *how much*. When asked how much wheat a farmer had, he didn't give a count of the number of grains. We use numbers for both of these, but in a very different way.

A third concept, very probably understood even by primitive peoples, possibly even by pre-humans, is that of *matching*. You don't need to actually count collections to determine whether they have the same size. If everyone in your hunting group holds a club then you know that you have *just as many* people as clubs. If there are still people without clubs you know you have *more* people than clubs. You don't need to know *how many* to know *just as many* or *more* or *less*. So matching is likely a more primitive idea even than simple counting.

Then, The Babylonians, Egyptians, and The Greeks

The ancient peoples in the historic era made great strides in two areas of mathematics: Geometry and Arithmetic, what we would today call Number Theory. Civilization, and agriculture, arose in the Fertile Crescent, today's Iraq, and spread from there. Agriculture meant that some people were freed of the task of actually finding their own food. Some of this excess potential went into mental activities such as philosophy, the sciences, and mathematics. The Babylonians created interesting number systems to help them count and measure. They developed the operations with numbers that we use today; multiplication and division, for example. The Egyptians, following them, were familiar with prime numbers. The Greek mathematicians raised Geometry to a high level and introduced Number Theory, including, for example, algorithms for finding the greatest common divisor of two integers.

And Elsewhere

While many people are most familiar with the heritage derived from the Greeks, there were parallel developments wherever civilization spread. China, India, and the Arab lands all had high civilizations and many people devoted to intellectual activity. Indeed, the numerals we used today have been inherited from Arabic mathematics. But such studies and advances were only known to a relatively few great thinkers. The same was true in the Americas as civilization grew, driven by agriculture.

Most of the mathematics done was very practical in nature. The advances were generated by the intellectual curiosity that drives mathematics forward. Hard problems and tricky questions led those with the necessary background to develop interesting and useful solutions. How much stone needs to be gathered to build a pyramid 200 cubits on a side and 122 high? Suppose we truncate it at 100 cubits high? Hmmm. Well, suppose it had a triangular base instead? Or circular?

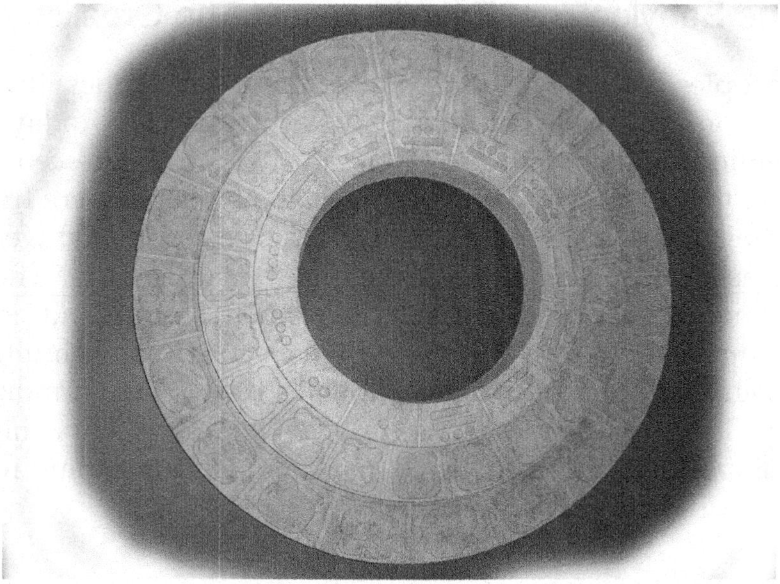

Maya Calendar

3 Natural Numbers and Axiom Systems

Before the 19^{h} century, mathematics was a bit ad–hoc. People discovered many things, to be sure, but it took a simple insight to begin the systematization of the study. The Greeks, at least, had discovered much earlier that one could derive some facts from other *simpler* ones using logic. This led people to wonder, what were the *simplest* ideas and could everything be derived from them.

Proof in Mathematics is a bit of a slippery thing. Most people think of it as a purely logical derivation, but in fact, it really means a demonstration that is acceptable to other mathematicians. Mostly that does mean logic nowadays, but old proofs seem a bit quaint now. If you wanted to convince a colleague of something you discovered, you would start with something he or she already knew (believed) and work from there. But there was a huge push in the 19th century to put math on a *solid footing*. Mathematicians today call this *rigor* and hold it in high regard.

Axioms

An *Axiom System* is nothing more than a set of statements that we take to be true for the purpose of seeing what can be derived from them. They are by definition *true in that system*. Whether they correspond to anything real is of no consequence. It is sort of like saying "What if these are true? What would then also be necessarily true?" Axioms are also called Postulates.

When they were first proposed the goal was to find an axiom system that could be used to derive the usual rules of arithmetic. There were a few tries, but the one accepted today is the one proposed by Giuseppe Peano (1858 - 1932). His five axioms can be used to derive a lot about the Natural Numbers: the numbers most people count with: 1, 2, 3, … .

It is, however, interesting that not everything can be derived from Peano's (or any similar) set of axioms. We will explore this later, but ask that you keep it in mind as we proceed.

We shall state a number of axioms and definitions based on them. We will then derive much of elementary mathematics from this simple base. One difficulty you may have in following it is that you know too much already. The *facts* that you know seem obvious to you, since you have long familiarity with them. In following the proofs you will be tempted to apply these facts when they have not yet been proven true. Try to avoid that or you may skip over something important in the development. Keep a skeptical mind.

Definitions and Properties

The axioms as shown below are close to Peano's originals. I'll number them and also give them names that I hope you will find clever, but not too clever.

The first of the axioms, named "one", simply states that 1 is a Natural Number or, equivalently, that there exists a Natural Number named 1. This statement introduces two ideas. First is the idea of a *set* of numbers, where "set" may be defined rigorously or not, and also the idea that 1 is a member of this set. We make two observations.

```
(1 = one) 1 is a Natural Number
```

First is that little can be derived from this single axiom. This is not surprising. Derivations involve implications of the *interactions* between axioms and with only one, there are no interactions. The second observation is that this current *system* (of a single Axiom) can be modeled in many ways. Any set with at least one element is a model, in fact, if you simply designate one of the elements to represent 1. But this idea of modeling is incredibly important as we shall see. Note that the empty set is *not* a model of this system. But your model doesn't need to have anything to do with what you normally think of as numbers. With this one axiom, a single child, for example, is a model, as it is a set of one element and you can designate him or her 1. Many people do this in fact.

We also note that 1 is not actually *defined* here, or anywhere. It is just something that *is* a Natural Number. It's properties are implicit in the axioms, but there is no actual definition. This is true of axiom systems in general. *Terms* are used and only have meaning within the system defined by the axioms.

Of course you may think of 1 as the smallest number. But ½ is a number (as we shall see later). Then you may counter that 1 is the smallest *whole* number. Hmm. What about zero? Well, how about the smallest *positive* whole number. This can go on for a while, but note that this depends on some things not defined. What is "smallest"? What is "positive"? We want a development in which the ideas aren't defined circularly.

The second Peano axiom, "successor", says that for any Natural Number, there is a *successor* element that is also a Natural Number. The successor of x can be denoted x', or s(x). We introduce such notations in mathematics simply to make it easier to speak and write about the concepts. It is a simple form of abstraction, and abstraction is one of the biggest ideas of all in math. It is also a shorthand, letting you say something meaningful in fewer words or symbols.

> (2 = successor) If x is a Natural Number, then there is a Natural Number, denoted x', called the *successor* of x.

Peano was trying to get at the idea that 2 is nothing more than the successor of 1, 3 is the successor of 2, etc. But with just two axioms, other systems can model it just as well. In fact a set with just one element is also a model if we designate the successor of 1 to be 1 itself. Both axioms are fulfilled. To rule out this possibility, we need more axioms. But again, the axiom doesn't *define* successor. It could be given any one of several interpretations. Note that we could use a different axiom here that specifies that the successor of an element is *distinct* from the element. We don't do that, since we will be able to derive that eventually from this and other axioms.

We shall see below that, as a consequence of another axiom, that the successor is unique. This rules out tree-like branching systems. But we can't conclude that quite yet. Note that *distinct* and *unique* are not the same idea.

If y is the successor of x, then x is called a *predecessor* of y. Later we will see that there can be at most one predecessor. In fact, x is the predecessor of y *only if* y is the successor of x. But we require additional axioms to be able to conclude that.

Another simple model of the system so far is the numbers on a clock. On a standard clock, the successor of 12 is 1 again, and round we go. We also need to rule out this possibility. Shown below is a "clock" with seven values. Here 1 is the successor of 7.

A Simple Seven Hour "Clock"

We can rule out both of these models (one is its own successor or one is the successor of something else) by adding a third axiom, "not one": The value 1 is not the successor of any Natural Number. We would like to think that this axiom, together with the others, implies that the set of Natural Numbers is infinite, but life, as math, isn't quite that simple. For example Consider the set {1, 2} where 2 is the successor of 1 and 2 is also the successor of 2. It satisfies all three axioms, but isn't what we want. We could either replace this axiom with another that can be modeled only by something resembling what we normally think of as the Natural Numbers, or we could just try to add more, as we shall.

> (3 = not one) 1 is not the successor of any Natural Number

This means, of course, that 1 has no predecessor.

But before we go on, we have to be a bit more careful. What does the third axiom really mean? How can we tell if $x' = 1$ for some x unless we know the meaning of *equals* itself? Yes, it really *does* depend on what the meaning of "is" is. For this we need a few axioms to define equality. But rather than trying to define it generally, we will define it only for the Natural Numbers. So here we have four more axioms, just to define equality, but also to put our axiom 3 on a firmer base. These weren't initially included in Peano's axioms, likely because they seemed to be obvious: shared knowledge among mathematicians.

First, we define *equality* as a relationship between pairs of Natural Numbers with the following properties: First $x = x$ for any Natural Number x. This is the *Reflexive Axiom*: a Natural Number is equal to itself. Don't just dismiss this as to trivial to talk about. If we want to put math on a solid footing, we need precision, rather than just depending on a shared but perhaps somewhat fuzzy understanding of the meanings of things. The naïve idea of "is the same as" is replaced by the precise notion of equality.

The second axiom of equality is that if $x = y$ for Natural Numbers x and y, then $y = x$ as well. This is the *Symmetric Axiom*. Note that not all relationships obey symmetry, or even reflexivity, for that matter. The less than relationship, defined later, obeys neither of these; $5 < 6$, but 6 is not less than 5 as well. Likewise 3 is not less than 3, so the reflexive property doesn't hold for $<$.

The third axiom defining equality is called the *Transitive Axiom*. If $x = y$ AND $y = z$, then $x = z$ as well. Note that this is an axiom, not something we prove. Not something we believe or not. It is just an axiom and the entire system of Natural Numbers can be derived from axioms like these (up to a point, remember). The final axiom is just that a Natural Number isn't equal to anything other than a Natural Number.

(reflexive) If x is a Natural Number, then $x = x$

(symmetric) If x and y are Natural Numbers and $x = y$, then $y = x$

(transitive) If x, y, and z are Natural Numbers, with $x = y$ and $y = z$, then $x = z$

(closure) If x is a Natural Number and $x = y$ then y is a Natural Number.

We can model the equality axioms in a set like {1, 2, 3} simply by declaring that none of them are equal to any others (except themselves, of course). So the entire relation here is just 1=1, 2=2, and 3=3; nothing else. We write $x \neq y$ as a shorthand for "it is not true that $x = y$". An alternate model would be to declare that each of them is equal to both of the others as well as itself. In this model, we would be faced with a dilemma. If we want to think of 1, 2, and 3 are distinct, but equal, then equals means something different from "is the same as". Does the set have one element or three? It all depends on the model. But a model in which every number is equal to every other isn't very useful in mathematics.

Continuing with our axioms of the Natural Numbers, proper, we come to one that does rule out clock-like systems: the *Injection Axiom*. This one is a bit more complicated. It has two parts. (a) If x and y are Natural Numbers and $x' = y'$ then $x = y$ as well. (b) Also, if $x = y$ then $x' = y'$. Note how we need equality here. But we have also introduced a conditional structure: if...then... We will leave that informal for now (as well as the word AND that we used above), but ideally this should be formalized as well.

(4 = injection) If x and y are Natural Numbers then $x' = y'$ if and only if $x = y$

Now, the system in which the set is {1, 2} and 2 is the successor of both 1 and 2 is ruled out, since the new axiom would imply that 1 = 2. But then the transitive axiom would imply the following: 1 = 2 = 2' and so 1 would be (=) the successor of 2, but this is ruled out by the third Peano axiom.

The construction "if and only if" recurs over and over in mathematics as in logic. "P if and only if Q" is really a shorthand for two statements: "if P then Q", and "if Q then P". An abbreviation is P iff Q.

We could expand the injection axiom to the two statements: *if* $x = y$ *then* $x' = y'$ and *if* $x' = y'$ *then* $x = y$.

Thus *injection* implies two things. First is that the successor of a Natural Number is unique. That is, if x had two successors they would need to be equal. Second that if we have two successors that turn out to be equal, then they are successors of the same Natural Number.

With this axiom, a clock-like system is ruled out since if, for example, 2 is the successor of 12 (1 can't be, of course), then 2 is the successor of both 1 and 12, so that 1 = 12. But then, presumably, 1 is the successor of 11, since 12 is, and we violate the third axiom again.

However, another model is not yet impossible. Imagine a set that contains two sequences, each like your normal idea of the Natural Numbers. In effect we have two things that each behave like 1 in their own sequence, with successors following them, but unconnected to the other sequence. Connections between the two sequences are ruled out by injection, but not if the two sequences are entirely distinct. Hmmm. What to do? The second sequence might even be clock-like.

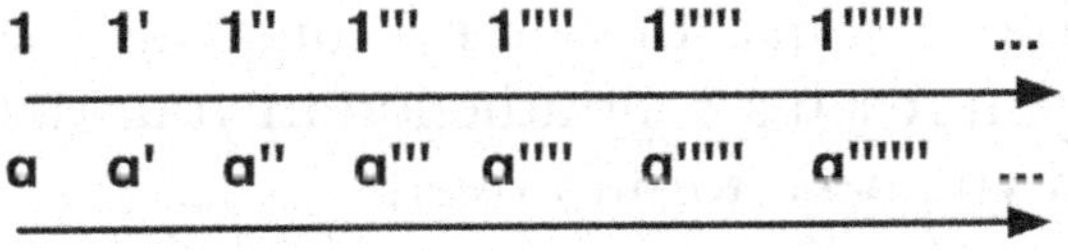

A Double Sequence

Note that symbols like 2, 3, 4, etc. are just convenient shorthand for 1', 1'', 1''', etc. They don't add anything new, but make it more convenient. So this double sequence idea could be modeled by having two symbols 1, and, say α, with successors as usual: 1, 1', 1'', along with α α', α'' and so on. There is nothing in the axioms to require this, but nothing yet to rule it out. What we really want is some way to say "all of the successive successors of 1 *and nothing else*." But we don't have that yet.

Before we present the final Peano axiom, I'd like to say a word about models. Something is a model if you can associate (map) the undefined terms in the axioms (like 1 and successor) to concepts in the model in such a way that the axioms become true. We have used models here only to show that the first few axioms alone don't really define what we are looking for. But models have a much more important function in axiomatic systems. If there is a "real world" model of an axiom system, then that system must be logically consistent since the world itself is. *Consistency* means that it isn't then possible to prove that something is both true and false using the axioms. Unfortunately, the infinite sequence model we have been using is not "real world" since the real world is (as far as we know) finite and this is not. So, these models aren't quite good enough to prove consistency, but they can be useful for our purposes here. An example of a real world model would be something like a particular flock of sheep as a model for a finite set.

The final axiom is very important if we want to try to make statements about *all* of the Natural Numbers and not just about finite subsets. Note that I can't write down an infinite sequence of statements, so if I want to say that something is true for all, I can't just start out: it is true for 1, it is true for 2, it is true for 3, etc. infinitely. We need a much more powerful tool to be able to make universal statements about the Natural Numbers.

Before stating the axiom, I again want to emphasize that we don't "prove" the axioms. We just accept (or reject) them as the basis of proving other statements. If a mathematician accepts these axioms, then they must also accept propositions derived from them. But if you reach a contradiction in your derivations then your axioms are inconsistent and you need to start over.

Suppose we have a statement P(*x*) that depends on a Natural Number *x*, but doesn't specify which. Such a statement could be true or false, depending on *x*. The statement "*x* is even" is true for some, but not all Natural Numbers. So is the statement "$x > 42$". This last axiom talks about such statements.

> (5 = induction) Let P(*x*) be a statement that depends on a Natural Number, *x*. If P*(1)* is true AND if, for every *x* for which P*(x)* is true, P*(x')* is also true, then P*(z)* is true for every Natural Number *z*.

This is a somewhat complex statement for a beginner. It is conditional on two parts (because of the AND). One of them is simple but the other harder to digest. It is just an axiom, however, and if we decide to include it it will let us make statements about *all* of the Natural Numbers. Not every mathematician accepts this axiom because it makes a claim about all the elements of an infinite set. They prefer to make only statements about finite things. If you reject it, the set of theorems you can prove is not the same as if you accept it, but that is the only thing you can say. This statement is no more "true" or "false" than any other axiom. If you accept it, you accept it as true. There are, in fact, simpler axioms that might rule out our double sequence model, but not be as *strong* as induction. As a matter of fact, exploring existing mathematical systems by *removing* axioms is a valid, and interesting method of inquiry. What can still be proven in the reduced system?

Note that this axiom rules out the model consisting of an infinite sequence starting with one together with a clock, starting with α. Consider the following. Say that a number can be *derived from 1* if it is either 1 or a number that is the successor of a number that can be derived from 1. Then, by this definition, 1 can be derived from 1. So can 2 since it is the successor of 1, which can be derived from 1. Note that the numbers in the clock part of the model can't be derived from 1. But, let's apply the fifth axiom (Induction) to this statement. The statement p that we want to use is: $P(y)$ = "*y can be derived from 1*". Let y be any number that can be derived from 1 and consider y'. Then y' is a number that can be derived from 1 since it is the successor of a value that itself can be derived from 1. But then, (since 1 can be derived from 1 and the successor of any number that can be derived from 1 can also be derived from 1), all of the Natural Numbers can be derived from 1. i.e. the statement "x can be derived from 1" is true for all Natural Numbers, x. So the values in the clock portion can't be Natural Numbers. The only remaining model is the single infinite sequence.

In the computer age, it has been more common to start with 0, rather than 1. Here we chose 1, since Peano did, but also because we want to say some things later about extending number systems and this will give an opportunity for a simple extension. But everything we have done so far works just as well starting at 0, or anywhere, really. Recall that the axiom system doesn't define what 1 is, though the properties of addition and multiplication, when defined, will give 0 and 1 a special place. For our purposes here, so far, we could just as well have started with –5.

The reason that 0 is important in the computing world is that starting our counts with 0 instead of 1 makes some internal computer representations as well as some programs a bit simpler.

However, as far as Peano's axioms are concerned the symbol 1 has no actual *meaning*. It is just a symbol for a concept (the starting point). It is given meaning from other axioms and definitions, such as the definition of addition.

A bit of terminology: The second part of the induction axiom is of the form "If $P(x)$ is true then $P(x')$ is also true." The IF clause here is often called the *induction hypothesis (or the induction assumption)*. We assume that it is true and use that to try to prove the *then* clause. We don't claim the *if* clause is always true, but only try to show that *when* it is true the *then* clause must also be true. This is the essence of the axiom.

Theorem 3.1: (Distinctness) If x is a Natural Number, then $x \neq x'$.

Proof: (By induction) Let P(x) be the statement of the theorem. We need to show two things. First, that P(1) is true and second that whenever P(x) is true, P(x') must also be true.

$1 \neq 1'$, since 1 isn't a successor, so P(1) is true.

Now suppose that P(x) is true for a Natural Number x. That is: $x \neq x'$ for a Natural Number x. We consider x' and must show that it isn't equal to *its* successor. But the injection axiom tells us that if $x' = x''$ it would imply that $x = x'$ which contradicts our hypothesis about x. Therefore P(x') is true. That establishes the theorem. QED

The last notation, Q.E.D., or QED, sometimes phrased in English as "quite easily done," is really a shorthand for the Latin: quod erat demonstrandum: "which had to be proven." Usually, however, the real meaning is ignored and this is just taken to mean "We are done, the theorem is proven, and if it has been intense to follow you can relax now for a bit."

We would like to show that the Natural Numbers form an infinite set. 1' = 2 (using the usual naming conventions), 2' = 3, 3' = 4, etc, but none of these can be equal to each other without violating the Injection axiom. The sequence can't "loop back" so it has to go on and on. Nor can we have other sequences that, like tree branches, "tap into" the normal ones as successors. We can't, for example have a situation like α, α', α'' being something new, but α''' being, say 5, because then α'' would have to be 4, etc. Finally, it is impossible for such a α''' to be, for example, 2 since then 1 would be a successor of something. That is because injection would force 1 to be equal to α' which is the successor of α. So, we know quite a lot about our number system, but obviously not everything.

What we have just given in the previous paragraph contains the germ of a *proof* of the *theorem* that the Natural Numbers form an infinite set, since *distinctness* holds. A theorem is just a fact that can be established from other facts in a chain of reduction ending in the axioms. We don't permit circular reasoning in math any more than we permit clock like systems in our model of the Natural Numbers. A proof is a demonstration, generally using logic, hopefully simple, that the theorem is true. Each step is laid out so that another person with some facility in logic can follow and verify the reasoning. The sequence of steps used in a proof can sometimes be used in proving other, similar, theorems. This results in a *proof method* and a new proof method is an especially valuable tool. My doctoral dissertation was once thought to contain such a new approach.

In the following theorem I'll use a proof method called *Proof by Contradiction*. I'll assume, or hypothesize, that the theorem is false and show that this would violate an axiom.

Proof by Contradiction: To show something in an axiomatic system is true, first assume it is false, or that its negation is true, and then show that some axiom must also be false. The negation of the statement is then inconsistent with the axioms. Since an axiom of logic is that something can't be simultaneously true and false, but must be one or the other, the original statement is true.

Often proofs are laid out very formally. Let's see what it might look like for the above.

Theorem 3.2: The Natural Numbers form an infinite set.

Proof: Assume that the statement of the theorem is false. If so, the Natural Numbers form a finite set. Then we have exactly n elements in the set of Natural Numbers for some value n. But, since each element has a successor, we must have n successors and they are all Natural Numbers by the successor axiom. But 1 isn't a successor so we must have n values different from 1 or at least n + 1 values overall unless two or more of the successors are equal to each other. But then we must have two *distinct* elements, say k and m, such that $k' = m'$. Injection then tells us that $k = m$, so they are not distinct after all. Therefore we have a contradiction and so the Natural Numbers form an infinite set. QED

Theorem 3.3: (Predecessor): If x is a Natural Number and $x \neq 1$, then x has a unique predecessor.

Proof: (By induction). Let the statement P(x) be the statement of the theorem. Then P(1) is true since the statement explicitly excludes 1 from consideration. We now must show that whenever P(x) is true, then P(x') is also true. Let x be a Natural

Number. Assume the theorem is true for x. Then, suppose $x \neq 1$, and x has a unique predecessor. Call it p. Then $p' = x$. But x has a successor x', so x' has at least one predecessor. We must show it is unique. Let q be any predecessor of x'. Then $q' = x'$ and so $q = x$ by injection. Thus x' has a unique predecessor. QED

Perhaps you noticed a flaw in the above "proof". I finessed the initial conditions for the induction to work. Surely the theorem is true for x = 1, as stated. But, if for some reason it failed for x = 2 then the induction chain would fail, since the proof for 1′ = 2 wasn't derived from that for 1. Re read the "proof" very carefully to note this. Fix the above proof by providing an explicit case for x = 2. Use the injection axiom. And note that this is a fairly common place for an induction proof to fail.

In case you are wondering, the name of the injection axiom will be explained in the next chapter. It comes from the fact that the successor relationship is a special kind of function.

The above is a sample of a general pattern for proofs using induction. In general they look like this:

Theorem–form:(Induction) Start with a statement about an arbitrary Natural Number that might be true or false: P(y).

(a) Prove that P(1) is true

(b) Assume that P(x) is true and prove that P(x') must be true no matter which x.

Note, and this is very important, that 1 is not excluded as a value of x in part b of the above.

Alternatively you can replace (b) with:

(b) Assume x is not 1 and that for the predecessor of x, say t, P(t) is true. Then show P(x) must be true.

If we accept the axiom of induction and can show parts (a) and (b) then we must accept that P(y) is true for every Natural Number y. We will see a lot of examples of this. Part (a) of such a proof is often called the *base case*. Part (b) is the *induction case*.

The induction axiom gives us a way to prove a large number of theorems about the Natural Numbers. However it also gives us a way to *define* things on the entire infinite set of Natural Numbers.

Addition

One of the useful ideas, which had been known for centuries before Peano, is simple addition. But, if we want to be able to derive everything from the axioms, we need to also see how addition fits into this scheme. We note that the intent of the successor relationship is similar to addition of 1, so we can begin our definition of addition with that. For every x, $x + 1$ is x'. So far so good. We could define $x + 2$ to be x'', etc. However, we can't write down an infinite sequence of definitions, so this doesn't get us to $x + y$ knowing only what we have of the current axioms. But here is a trick. Define $x + y$ like this: If $y = 1$ then $x + y$ is x'. Otherwise, let z be the predecessor of y (i.e. $z' = y$) and define $x + y$ to be $(x + z)'$. The above theorem says that z must exist.

> $x + 1$ is defined to be x'
>
> If $y \neq 1$ then $x + y$ is defined to be $(x + z)'$ where $z' = y$.

Another way to state the second rule, is that $x + z' = (x + z)'$.

Now we need to consider is whether $x + y$ is actually defined. We can prove that it is, using induction. Here the word "defined" means that the $x + y$ is a unique Natural Number whenever x and y are Natural Numbers.

If $x + y$ is defined we call it the *sum* of x and y.

Theorem 3.4: Addition is well-defined: The above definition uniquely defines $x + y$ for all Natural Numbers x and y.

Proof: (using induction on y) Let x be any Natural Number. Statement P(y) is that $x + y$ is defined and unique for the Natural Number y.

(a) Let x be any Natural Number. Then $x + 1$ is defined (as x', which is unique by injection). So P(1) is true.

(*b*) Now let z be any Natural Number, for which $x + z$ is defined and unique. i.e. P(z) is true). Then $x + z'$ is defined as $(x + z)'$, the successor of $(x+z)$. Likewise, $(x + z)'$ is unique by injection, so if $t = (x + z)'$ then (*i*), $t \neq 1$, since it is a successor, and (*ii*) the predecessor of t is $x + z$ which is unique. So P(z') is true. Therefore P(y) is true for every Natural Number, y. Thus by induction $x + y$ is defined and unique for every Natural Number, y. QED

Accepting the Axiom of Induction, requires that you also accept a very large number of other statements about the set of Natural Numbers. But not all, of course. You can't use it to show that all Natural Numbers are even, since 1 isn't. Another reason you can't use it to show that all Natural Numbers are even is that 2 is even but its successor isn't.

We also note that there is an alternate, equivalent, formulation of the Axiom of Induction.

> If A is a set such that
>
> A contains 1, AND
>
> A contains the successor of every element in A
>
> then A contains all of the Natural Numbers.

I will normally prefer the first formulation, since this one requires some further work on the nature of sets that I will leave until later.

There is a relationship between these two formulations, of course. If we let A be a set and let P(x) be the proposition that x is an element of A then we see that the two formulations say the same thing as long as we have a good basis for the "element of" relationship. See how tangled up mathematics gets?

A third formulation is often used. It is equivalent to the others, but it takes a bit of work to show it. This is usually called the *Axiom of Strong Induction*. We won't use this formulation later and so omit the proof of equivalence.

> If P(1) is true AND if, for every x for which P(y) is true for all y from 1 to x, P(x') is also true, then P(z) is true for every Natural Number z

There is a danger here, that you might think that *anything* can be proved with induction. This isn't the case, of course, so it may help to see a situation, that is a bit ridiculous, but in which a proof by induction seems to work, until you examine it closely.

Theorem? (not): Any collection of coins if laid on a table will always show the same face, either heads or tails: P(n): In any set of n coins, all show the same face; heads or tails.

Proof: by induction on the size n of the set.

(a) P(1) is true as the single coin is either heads or tails.

(*b*)Assume P(n) is true for some n (and show P(n') is true). Assume we have a set of n' coins and drop them onto a table. Remove one, but don't turn it over. We now have a set of n and so all show the same face by the induction hypothesis. Now simultaneously return the first coin without changing its orientation and remove another, different, coin, without turning it over. Again we have n coins and so all show the same face. Now, return the second coin without turning it over. Since none of the coins were flipped, we now have n' coins all showing the same face. Theorem proved. QED

Note that the induction hypothesis, P(n), says that in any set of n coins they all show the same face. We used this twice above. Note that it is possible to pick a different coin in the above, since there are at least 2, since we proved the case for 1 separately.

Coins

You may think this is a false theorem, and so you must think there is a flaw in the proof, as their surely is. I'll give several hints about the flaw later, but it is worth thinking about why the "proof" is in error. Note that the same proof (in essence) has long been used to "prove" that "All horses have the same color."

To guide your thinking on this, note that the induction step in an induction proof really says that to show P(n) is true for all n, we need to show that whenever P(n) is true, P(n') is also true. The most important word here is *whenever*. Said another way, in *every* situation in which P(n) is true, P(n') is also true.

To show that something is true is often quite difficult. To show that it is false, however, can sometimes, at least, be quite simple. One way is to have a single *counterexample* to a general statement. Mathematicians often look for counterexamples when exploring whether something is true or not. Something can't *always* be true if it is false even in a single instance. But, that won't settle why we went wrong in the above "proof." Give it some thought. I've given a hint above, actually, and will give one or two others later in the book rather than spoil your fun with the real reason.

Let's note something simple, but deep. Every time we added a new axiom to our list we decreased the number of things that *could* be true, but increased the number of things that *must* be true (unless you add an axiom that contradicts the others in some way). So, some models that work with a small set of axioms are ruled out when we add another. But we gain the ability to prove things with a new axiom. That is, we show things that now *must* be true with those axioms. Ideally we would like to keep increasing the number of axioms until we get *exactly* what we want, a set of axioms that describes exactly the Natural Numbers. But it was shown in the previous century that this is impossible. In any consistent, nontrivial axiom system powerful enough to describe ordinary arithmetic, any well defined process to verify theorems won't be able to verify them all. Said another way, there will be statements that the process can't determine the validity of. The discovery of that was true mathematics. The perpetrator was Kurt Gödel, and it caused quite a bit of anguish back in 1931.

Before we proceed let's take a closer look at the IF–THEN statement that is so essential in the induction axiom. It has a precise meaning to a mathematician. If I say something like "If the sky is cloudy, then it will rain," I might be making a true statement or a false one. There is so much variability in the weather that this statement is unlikely to be true in general. Certainly you have seen many counterexamples. But, suppose, just for fun, that we lived in a world in which this is a true statement. According to the meaning of IF–THEN, it doesn't mean that it will necessarily rain. Maybe the sky isn't cloudy. It will be helpful to name some things, as mathematicians like to do. The part of the above IF–THEN "the sky is cloudy" is called the *antecedent*. The other part, "it will rain," is called the *consequent*. The *meaning* of an IF–THEN is that when the IF–THEN itself is true and the antecedent is also true, then the consequent must be true as well. But if you examine this carefully, note that it tells us nothing about the case when the IF–THEN is true but the antecedent is false. In that case, anything can happen. In other words, it gives us no information about whether it will rain or not. If the sky is blue (i.e. not cloudy) and it rains, we can't claim that the IF–THEN is false.

However, if the antecedent is true, but the consequent is false, then the IF–THEN itself must be false, or at least, not true in general. If both antecedent and consequent are false, the IF-THEN can still be true. Make sure that you understand this. It will be discussed again in the next chapter.

Properties of Addition – Proof Based on Axioms

We have a simple definition of addition. But that operation has a lot of well-known properties. We would like to see if our definition is adequate by trying to derive those properties, using nothing but the axioms and the definition. We shall see that the axiom of induction is essential in all of these. First we look at a few example proofs and then step back to get some perspective.

Recall that the induction axiom requires us to do two things to prove that a statement P(x) is true for every Natural Number x. Show that P(1) is true and show that if P(t) is true then P(t') is also true. If we can do that then P is true for every Natural Number.

The Associative Law

The Associative Law of Addition says that for all Natural Numbers: $(x + y) + z = x + (y + z)$

The proof will be by induction on z. Note that the proof is independent of the choice of x and y, so we don't need to do a double or triple induction over all terms.

Theorem 3.5: Associative Law of Addition: If x, y, and z are Natural Numbers: $(x + y) + z = x + (y + z)$.

Proof: (by induction on z) The statement of the theorem is $P(z)$.

(a) If z is 1 the left side is $(x + y) + 1$ which is $(x + y)'$ by the definition of addition. The latter is $x + y'$, also by the definition of addition. But this is just $x + (y + 1)$. So the base case is proved.

(*b*) For the inductive step we assume the theorem is true for m: $(x + y) + m = x + (y + m)$ and show it is also true for m'. The left side would be $(x + y) + m'$ which is $((x + y) + m)'$ by the definition of addition.

But by the induction assumption this latter expression is $(x + (y + m))'$.

Applying the definition of addition again we get $x + (y + m)'$, and again, $x + (y + m')$. So the theorem holds in the induction case as well. So it is proved for all Natural Numbers. QED

Because of the associative law we frequently omit the parentheses since it doesn't matter where we place them as long as it is correct: $x + y + z$.

The Law of Substitution

Theorem 3.6: The Substitution Law of Addition: If a, b, and x are Natural Numbers and $a = b$, then $a + x = b + x$.

Proof: Exercise. Use Induction on x.

This is called the substitution law, since if you have any formula involving addition and one of its terms, a, is equal to something else, b, then you can substitute b into that formula for a and have something equal to the original. Each operator that we will introduce, multiplication, for example, requires its own law of substitution.

However, the law of substitution also tells us that if we have two values that are known to be equal and we add the same thing to both, the results are always equal. This idea is often used in manipulating equations.

Exercise: If $a = b$ then how many times do we need to apply Theorem 3.6 to show that $(a + x) + y = (b + x) + y$? Assume a, b, x, and y are Natural Numbers. Note that this question has several correct answers. Say why you came to the conclusion you did. Even better. Say why, and how, the various correct solutions are arrived at.

The Commutative Law

The Commutative Law of addition says: $x + y = y + x$.

Theorem 3.7: Commutative Law of Addition: For any Natural Numbers x and y, $x + y = y + x$.

Proof: (three cases)

Case 1. Suppose both x and y are 1. The both the left and right sides are identical $(1 + 1 = 1 + 1)$, so they are equal. This completes the base case.

Case 2. Suppose y is 1, but x may not be. We want to show that $x + 1 = 1 + x$ for any x. Said another way this is $x' = 1 + x$ for any x. We will apply induction, with that as the proposition $P(x)$.

$P(x)$: $x' = 1 + x$ for any x.

The base case is just Case 1, above. For the induction case, assume that $t + 1 = 1 + t$ for some t. We consider t'.

We already know that $t' = t + 1$ from the definition of +.

So, since successors are unique, $(t')' = (t + 1)' = (1 + t)'$. The second equality is just the induction assumption.

We also have, from the definition of +: $1 + t' = (1 + t)'$.

Thus, putting the two above equations together, $1 + t' = (1 + t)' = (t')' = t' + 1$.

Case 3. Suppose, finally, y is not 1. Then y has a unique predecessor, say t.

Now assume the formula above is true for t: $x + t = t + x$: the induction hypothesis. We must show it true for $t' = y$.

First, $x + y = x + t' = (x + t)'$, by the definition of +.

By substitution, using the induction hypothesis, $(t + x)' = (x + t)'$ and so $x + y = (t + x)'$.

We want to show that the right hand side of this, $(t + x)'$, is the same as $t' + x$, which is $y + x$. So, it remains to show that $(t + x)' = t' + x$. Once this is done, the commutative law is proven. We again use the axiom of induction along with the definition of addition. Here, the proposition is:

$Q(x)$: $(t + x)' = t' + x$, for any Natural Number x. We use induction on x.

First, suppose x is 1. We have $(t + 1)'$ on the left, which is just t''. But the right side is $(t' + 1)$ which is also t''. So $Q(x)$ is true in the case $x = 1$. The base case is established.

Next, suppose x is not 1 but the formula is true for the predecessor of x, say z: $(t + z)' = t' + z$ and $z' = x$; We want to establish the truth of $Q(x)$.

Again starting with the left side of the formula: $(t + x)' = (t + z')'$.

Also, $\mathbf{(t + x)'} = (t + z')' = ((t + z)')'$ by the definition of +.

But $((t + z)')'$ is $((t' + z))'$ by the induction hypothesis and substitution.

One set of parentheses is redundant in the right hand expression, so we have just $(t' + z)'$ which is $t' + z'$ by the definition of addition.

But $t' + z'$ is $\mathbf{t' + x}$. So, following the chain of equalities, $Q(x)$ is true and the commutative law is proven. QED

Breaking a proof into cases is a common technique in mathematics. A proof by induction always has at least two cases, of course. There are some theorems that have proofs with a huge number of cases. Sometimes a computer can be used, either to enumerate all of the cases possible, or to provide proofs for some of the cases discovered otherwise. Such proofs aren't very satisfying to a mathematician. We seek elegance in our work, especially in our proofs. One classic problem, the Four Color Problem(+), was an open question for over a hundred years: is it possible to color any planar map with only four different colors? The answer was finally provided in 1976, so we now have the Four Color *Theorem* in Topology. The proof that four colors is sufficient has nearly 2000 cases and the cases themselves were examined and settled by a computer program. It is nice to have the answer, especially since many, many incorrect "proofs" were published over the years, but it isn't a satisfying proof. The division into cases was done by two mathematicians (Kenneth Appel and Wolfgang Haken) and requires several hundred pages.

What is Mathematics: A First Look

Mathematics is not just a collection of facts, though early mathematics education seems to give the impression that it is. Students are asked to memorize a lot of things. Teachers hope that these facts will make other things easier. One of the key parts of mathematics (as of language) is the idea of *abstraction.* An abstraction is a *name* for a, possibly complex, phenomenon in the world. A name is a simple thing, but as an abstraction it *captures* the complexity of the real thing. *Animal* is an abstraction. Animals, as such, don't exist in the world. Individual people and tigers do. The word *animal* has a meaning that captures part of the *meaning* of a person or a tiger. Abstractions can be directly based on things or they can be based on other abstractions. *Number* is an abstraction. *Natural Number* is an abstraction based on number. In this case, it is a more specific abstraction, just as *mammal* is more specific, but not more concrete, than *animal.*

So, one partial definition of mathematics is that it is the *art of abstraction.* Some of the abstractions are motivated by the things that occur in the world, but they need not be. Math is "pure mind stuff," invented by humans. It doesn't exist in the world. In this way, mathematics is not like physics or any science, which depend fundamentally on the real world. Some math is inspired by the world, and some math has application in the world, but it is really from a different domain.

Later we will ask the question again, and see some other aspects, including the central place of *proof* in mathematics.

If mathematics is about abstraction, then mathematicians are people who *create* abstractions and derive their properties. Mathematics is a *creative process* like poetry, not a fixed body of knowledge.

It is time to return for a moment to the false theorem we saw earlier. It must be false, of course, since I showed a counterexample. But I also note that in the "proof by induction" I lied. See if you can find the lie. Mathematicians aren't supposed to lie. Very bad form. Propaganda isn't mathematics, nor mathematics, propaganda.

The Law of Cancellation

With the help of the Associative and Commutative laws there are quite a few other things we might prove.

We showed earlier that $m + 1 \neq m$. But we can say more:

Theorem 3.8: If x and y are Natural Numbers, then $x + y \neq x$.

Proof: by induction on x.

(a) The case for $x = 1$ was proven above, as noted.

(*b*) Let x be a Natural Number for which the theorem is true: $x + m \neq x$ for any Natural Number m. Now consider x'.

Seeking a contradiction, suppose that $x' + m = x'$.

Then $x' = x' + m = m + x' = (m + x)'$. Note the use of the commutative law.

But the injection axiom then says that $x = m + x$ contradicting the induction hypothesis. QED

The theorem we proved just above starts mathematicians to wonder. What if there *were* a number k such that $x + k = x$? It couldn't be a Natural Number, of course, so we would need to extend our number system to make this possible. We consider this in Chapter 5. But now we have a glimpse into the mind of a mathematician. We are people who *wonder* about things. What if...? How could it happen that...? This is the tickle that gets our blood moving. What would make … true?

Theorem 3.9: Law of Cancellation. If x, a, and b are Natural Numbers and $x + a = x + b$ then $a = b$.

Proof: By induction on x. if $x = 1$ then $1 + a = 1 + b$ just says, using the commutative law, that $a' = b'$ so $a = b$ by injection.

Let m be a Natural Number with the property that for Natural Numbers a, and b, when $m + a = m + b$ it is also true that $a = b$. Consider m'. Does $m' + a = m' + b$ require $a = b$? Well, $m' + b = m' + a = (m + 1) + a = m + (1 + a)$. Likewise $m' + b = (m + 1) + b = m + (1 + b)$. So, $m + (1 + a) = m + (1 + b)$. But $1 + a$ and $1 + b$ are Natural Numbers of course, so the induction hypothesis implies that $1 + a = 1 + b$ and again, injection requires $a = b$. QED

Note that a few steps were omitted from this proof. Can you name the axioms and theorems needed to complete the steps? Theorem 3.9 is called the *cancellation law of addition*. Compare it with the law of substitution. Logically they are called *converses* of each other. The converse of a statement "If X then Y" is the statement "If Y then X". Depending on X and Y, a statement and its converse can be both true, both false, or either one true and the other false. As an exercise it would be worth while to find examples of each of these four cases.

Summary So Far

We have a set, the Natural Numbers, and one operation, addition. We also have one relation, equality. The elements of the set have certain properties as does the operation. We could do more at this point, such as multiplication, but will delay this until we first develop an extension of the Natural Numbers.

It is important to note that axiom systems in general are very important in modern mathematics. The Peano Axioms are sort of a model for the general situation, but there are many others. There are many fields such as Abstract Algebra and Topology that generate complex structures from a few axioms and result in systems that are not much like number systems at all.

An axiom system, while it doesn't define its primitive terms, is, in fact, a definition of some structure. But the structure my not be obvious at first and requires elaboration and exploration. Definitions formulate ideas within the axiom system and theorems state conclusions that we can make of the consequences of the axioms and definitions.

When we state an axiom or a definition in mathematics, we try to say as little as possible. For example, the definition of addition told us how to add 1 on the right hand side of a sum, but not the left. Saying too much, by including addition on the left as well, seems like a good idea until you realize that this might introduce an inconsistency into the system. Simpler is better. Of course we then needed to prove that it works from the left as well using one or more theorems. But a system with inconsistencies is useless. We would need some proof of consistency. Suppose we could arrive at a formula from two different paths in which the two paths implied different values for the formula.

Theorems and proofs are the breadcrumb trail that mathematicians leave back to home base: the axioms. It is useful to name the important theorems to aid communication among mathematicians. The cancellation law, above, is an example. Proofs are like the rocks that you use as stepping stones to get you across a stream, establishing an extension of the known world.

Just as we like to keep statements of axioms and definitions simple we also like to be careful about how we state theorems. If a theorem has the form "If X, then Y", we like to say as little as possible in the X statement, called the *hypothesis*. If we say more than we need to say then that lessens the applicability of the theorem. Suppose, for example we have two specific propositions P and Q and a third R. Suppose we say something like "When P and Q are true, then R must also be true." But suppose that, for whatever reason we don't really need Q in any way to establish R and we could have stated it without reference to Q. Suppose also that P and Q are such that they can be independently true and neither depends on the other. Then a mathematician in a situation in which P is known to be true, but not Q, can't conclude from the theorem that R is true. So, if you are writing theorems, say as little as possible in the hypotheses of the theorem. But include what is necessary, of course.

4 An Introduction To Sets And A Bit Of Logic

Sets – Definitions and Properties

The idea of a *set* is seemingly simple, but it hides many difficulties for mathematicians. Here I'll say a few things about naïve set theory and even give a few axioms, but won't try to give a complete definition. If I just use common language, it is hard to get a grasp on what a set is that doesn't involve circular reasoning. A set is a collection of objects. What is a collection? Well, a set. Round and round. An axiomatic approach seems simpler since we needn't define the primitive notions and just make sure we have something consistent with common models of sets.

Sets are "defined" by their elements. If A is a set and m is an element of A we write $m \in A$. We also say that m is *in* A. Some set A is called a *subset* of B, written $A \subseteq B$ provided every element of A is an element of B. Two sets A and B are equal provided $A \subseteq B$ and $B \subseteq A$.

Moreover, an element x is either in a set A ($x \in A$) or it is not ($x \notin A$). It can't be both and it can't be neither.

If $A \subseteq B$ we also write $B \supseteq A$ and say that B is a *superset* of A.

If each of two sets A and *B* is a subset of the other then they are *equal*. They have exactly the same elements.

A set A is a *proper subset* of a set B if $A \subseteq B$ but $A \neq B$. This implies that B has an element that is not an element of A.

While this doesn't *define* either *set* or *element*, it does require that elements be *distinguishable*. There needs to be some way, such as an equality relation, to determine if two elements are the same. Oddly, this can be surprisingly subtle.

However, in reality, these are just informal statements that might become axioms in a careful development. A possible interpretation might apply the statements to

Natural Numbers. The elements are Natural Numbers and the sets are arbitrary collections, such as {2, 3, 5, 7, 11} or the even numbers. One way to specify the elements of a small finite set is to list the elements between braces as we have just seen. For large sets this isn't feasible, of course.

The *union* of two sets, written $A \cup B$ is the set of all elements that are in either A or B or both. The *intersection*, $A \cap B$ is the set of all elements in both sets, A and B. The union of {1, 3, 5, 7, 9} and {2, 3, 5, 7, 11} is {1, 2, 3, 5, 7, 9, 11}. The intersection of those two sets is {3, 5, 7}.

There is a set with no elements, called the *empty*, or *null*, set, denoted $\varnothing$.

Note that $X \cap \varnothing = \varnothing$ for any set X. Also $X \cup \varnothing = X$ for any set X.

Two sets X and Y are called *disjoint* if they have no common elements. This is the same as saying $X \cap Y = \varnothing$.

If a is an element then there is a set containing only a, denoted $\{a\}$. Such a set is called a *singleton*. Note that x is distinct from $\{x\}$. If a and b are distinct elements then there is a set consisting of a and b and nothing else, denoted $\{a, b\}$, or equivalently, $\{b, a\}$.

As you can see, we often give names to sets. A common name given to the Natural Numbers is $\mathfrak{N}$. This is just N written in a Germanic font. Writing $x \in \mathfrak{N}$ says that x is a Natural Number.

Sometimes we need a way to specify a set (such as a set of numbers) without listing all of its elements. In general, $\{x \mid P(x)\}$, where P is a proposition that can be true or false, is the set of elements, x, for which $P(x)$ is true. This is sometimes called *set builder notation*. For example $\{ x \mid x > 5 \text{ and } x < 10 \}$, assuming we are restricting the domain of discussion to the Natural Numbers is just a way to specify the set {6, 7, 8, 9}. Likewise $\{y \mid y > 42\}$ is the set of all (natural) numbers greater than 42; an infinite set. Inside braces we start by listing a variable name followed by a vertical stroke. After the stroke we list a set of conditions that determine if a value for x is in the set or not. The last example above is read, *the set of all y such that $y > 42$*. Note that if we use x instead of y in the definition it makes no difference. In due course we will define the relations < and >.

The *size,* or *cardinality,* of a set is the number of distinct elements it contains. A singleton has cardinality 1. The empty set has size 0. Some sets have infinite cardinality and we will discuss this later. Infinity is a complex topic.

Now I'll step onto some dangerous ground. Suppose P(x) is a proposition that can be true or false for an element x. And suppose that P somehow provides an effective way to determine the truth. For example P(x) = "x is a Natural Number." Another is P(y) = "$y > 42$." The effective procedure implied in the first example is the Peano Axioms. But Q(x) = "x is a red horse," probably isn't the sort of proposition that we can use, since "red horse" is too subjective. But supposing that P is such a "nice" proposition then there is a set containing exactly all the elements for which P is true. { x | P(x) is true}. This set might also be called P if we can avoid confusion.

The reason why this is dangerous ground is that mathematicians have often tried to play with the "set" { x | x is a set}, or the set of all sets. The reason why the proposition doesn't work is that there is really no *effective* procedure for determining the truth of "x is a set." Remember that an axiom system leaves its primitive terms undefined. The definition, such as it is, is an elaboration of the axioms through the theorems, not an effective procedure except in certain circumstances. The "set of all sets" idea leads to paradox.

There is, however, a way to explore things more general than sets. A *class*(+) is a collection of sets that can be defined precisely by a proposition. A class isn't necessarily a set, however. So you can talk about the *class* of all sets, but not the *set* of all sets. The rules for manipulation of classes are different from that for sets. We leave it to your future investigations.

We do, however, want to talk about sets of sets. For example, {{0}, {1, 2}} is a set, but its elements are sets {0}, and {1, 2}. So { x | x is a set of Natural Numbers} is perfectly allowable. If we agree that we include the empty set in this, then it is called the *power set* of the Natural Numbers. But note that this set contains no Natural Numbers, just sets of Natural Numbers, like {2, 4, 6}, or even {x | x is an even Natural Number}. If a set X has n elements, then the power set has 2^n elements. For this reason the power set of X is often denoted 2^X

In particular, we want to talk about the *complement* of a set X: the set of elements that are not in X. Again, there are too many possibilities unless you limit it. If you are talking about sets of Natural Numbers, then you can talk about the complement of a set within this context. Not "everything not in X", but every Natural Number not in X. "Everything" is just too general.

Another shorthand helps us express long sums. For example: $\sum\{\, x \mid x > 5$ and $x < 10 \,\}$ is just $6 + 7 + 8 + 9 = 30$. The symbol $\sum$ represents the sum of all elements in a set. Read the example as "The sum of all values, x, such that $x > 5$ and $x < 10$." The vertical stroke is "such that". We could add, "x is Natural Number and" after the stroke to be more explicit, or we could just understand it as part of the domain of discourse. If it could be interpreted ambiguously, then it is best to add it: $\sum\{\, x \mid x \in \mathfrak{N}$ and $x > 5$ and $x < 10 \,\}$. An alternate notation for this is: $\sum_{x>5}^{x<10} x$

Definition 4.1: If A and B are two sets, then $A \times B = \{\, (a, b) \mid a$ is in A and b is in B$\}$ is called the *Cartesian Product* of A and B, named for René Descartes (1596 – 1650). The notation is read as "the set of *all* ordered pairs (a, b) *such that* a is in A and b is in B." So $\mathfrak{N}^2 = \mathfrak{N} \times \mathfrak{N}$ is just the Cartesian Product of $\mathfrak{N}$ with itself. Note that, even when the two sets are the same, as in $\mathfrak{N}^2$, the pair (x, y) is not the same pair as (y, x), unless $x = y$, hence the name "ordered" pair.

It is possible to give a rigorous definition of the idea of "ordered" pair. We can specify the "first" element explicitly. $(x, y) = \{\{x, y\}, x\}$. So it is a set consisting of a set of two items and, separately, one of those items to be designated as the first element. Then $(5, 2) = \{\{5, 2\}, 5\} = \{\{2, 5\}, 5\}$. But $\{\{5, 2\}, 3\}$ doesn't represent any ordered pair since 3 is not an element of $\{5, 2\}$.

Definition 4.2: A *relation, R,* between sets S and T is an arbitrary subset of $S \times T$. The first element of an ordered pair in R is said to be *related to* the second. If $(s, t) \in R$, we write sRt. We often use symbols for relations, such as = or <. Then we write $x = y$ instead of saying that "(x, y) is an element of the relation =."

Definition 4.3: The *inverse* of a relation R in $A \times B$ is formed from the set of ordered pairs of R by reversing the components. It is a relation between B and A. If R is a relation between A and B, then the inverse $R^{-1} = \{(b, a) \mid (a, b)$ is in R$\}$. It is a relation in $B \times A$.

The superscript –1 is often used in mathematics to indicate an *inverse* of some sort.

Definition 4.4: A *function from A to B* is a relation in A × B in which every element of A appears exactly once as a first element. Set A is called the *domain* and B the *range* of the function. If f is a function from A to B we write f:A→B. If x is an element of the domain of a function f, then f(x) denotes the corresponding element of the range. Namely, (x, f(x)) ∈ f. Another, equivalent, characterization of a function is: a relation f is a function provided that if (a, b) ∈ f and (a, c) ∈ f, then $b = c$.

The successor relation in 𝔑 × 𝔑 is a function. Every Natural Number has exactly one successor. Predecessor is not a function in 𝔑 × 𝔑, since 1 has no predecessor.

The operators, such as addition and multiplication are also functions. The domain of each is the Cartesian product of the Natural Numbers and itself, so the range is also the Natural Numbers. +:𝔑 × 𝔑→𝔑. For example, +((3, 4)) = 7.

Functions from a set A to a set B are, I hope, familiar to you. They are often given by a formula that gives the second element in terms of the first via some computation. This assumes that A and B are similar in some sense. For example, the function f(x) = 2 · x, can be interpreted as a function from the Natural Numbers to the Natural Numbers in which the second element is twice the first. That is to say, the ordered pairs in the function are of the form (x, f(x)), or (x, 2·x).

If we use a notation such as f(x) = y, where y just stands for the computed value of the function for a particular value of x, we call x the *independent variable* and y the *dependent variable*. In other words, the value of y (i.e. f{x)) depends on x.

Definition 4.5: A function f from A to B is said to be *onto*, provided that every element of B appears as a second element *at least once*. Another way to say this is that if y is an arbitrary element in B then there is at least one element x in A such that $f(x) = y$. Onto functions are also called *surjections*.

Note: The inverse of a function from A to B is a relation, but it need not be a function. It might not even have an element (b, x) for every b in B. If the original function is a surjection, however, it will have such an element, possibly several.

The function above (f(x) = 2 · x) is not onto the Natural Numbers, since all of the second elements of the ordered pairs are even, so we say it is *into*, instead. But it is onto the set of even Natural Numbers.

Definition 4.6: A function from A to B is said to be *one-to-one* (or 1–1) provided that an element of B appears *at most once* as a second element. An alternate name for a 1–1 function is *injection*. Injections have the property that if $f(x) = y$ and $f(z) = y$ then $x = z$.

Peano Axiom 4, injection, gets its name from this definition since the function from a Natural Number to its successor is an injection.

The doubling function, above, is 1–1. A given even number is the *image* of at most one Natural Number using our example function. The value f(3) = 6 but no other Natural Number yields 6 when doubled. The absolute value function (defined below) is not. Both -3 and 3 have the same absolute value. But we need to define things like –3 before we can get there, of course.

Definition 4.7: A *bijection* from A to B is an injection from A onto B. That is to say there is a 1–1 mapping from A onto all of B. It is both an injection and a surjection.

Theorem 4.8: The inverse of a bijection is a bijection.

Proof: The inverse is a function, since each element of the range will be in the first position of the relation exactly once. The inverse is onto by definition of a function. The inverse is 1–1 since the original was a function. QED

If C is a subset of A and f is a relation or a function from A to B, then $f(C) = \{f(x) \mid x \in C\}$. The set f(C) is called the *image* of C in B.

Likewise if D is a subset of B then $f^{-1}(D) = \{x \in A \mid f(x) \in D\}$. The set $f^{-1}(D)$ is called the *inverse image* of D in A.

Here is a silly example to illustrate these definitions. Suppose I have a flock of sheep and a bunch of numbered bells. Suppose the bells are numbered 1, 2, 3, etc. using successive Natural Numbers up to some maximum.

Suppose I want to put bells on the sheep. If I put one bell on each sheep and I run out of bells exactly when each sheep has one bell, then I have a bijection between the bells (and their numbers) and the sheep. The number on the last bell tells me how many sheep there are of course.

If I run out of bells before I run out of sheep then I have an injection but it isn't "onto" the set of sheep, so isn't a bijection.

If I run out of sheep before I run out of bells then I have a relation but not a function, since not every element of the domain (bells) is matched with a sheep.

If I put several bells on some of the sheep (at least one), then it isn't an injection, though it may be a surjection (if each sheep gets at least one bell) or not (if some sheep are left without bells). It will still be a relation, however, and may be a function, provided I use all the bells. So one sheep may get bell 3 and also bell 5.

But if it were possible to put the same bell on two or more sheep (pity the sheep) then it wouldn't be a function either. But, yes, still a relation.

The inverse of any of these is some sort of matching of sheep–to–bells, rather than bells–to–sheep as described above. Note that the inverse of the relation could be a function even if the bells–to–sheep relation isn't.

There, I'm sure that cleared it up.

Theorem: Mathematics is all about sheep.

Proof: False: Counterexample. Sometimes it's about bells, or coins, or ...

Simple Logic

Logic is about correct reasoning. Logic is a large topic and there are several varieties, from Aristotelian Logic to Mathematical Logic. Here I will introduce only enough of logic to show how we prove most theorems. A much more complete study is needed to advance far in either mathematics, computer science, or philosophy, for that matter.

Many theorems have the form "If P, then Q" where P and Q are *propositions* of some kind. A proposition involves the relationship between two *terms*. A term represents some *thing* and the relationship may be true or false. "Mary is a woman", is a proposition with terms *Mary* and *woman*. It might be false. Mary could, in fact, be a puppy. Just because we state a proposition we aren't necessarily asserting that it is true. The theorem form we have indicated above, "If P, then Q," is called a *conditional*. It can also be true or false. We take it to be true provided that whenever the proposition P is true the proposition Q is also true. "If Mary is a woman, then Mary is tall" is likely false, though that depends on the domain of discourse. However, we also take the conditional to be true whenever P is false. If Mary is actually a puppy then "If Mary is a woman then Mary is tall" is taken to be true, even if Mary is a short puppy. A simple conditional is "If it is red then it is an apple." This might be true or false, depending on context. Normally you would probably take it to be false, but if the set of choices were restricted, the context, it might be true. If your set of choices were just apples and kiwi fruits it might still be false, since some apples are green, like kiwis.

In a conditional statement, "If P, then Q," the proposition P is called the *antecedent*, and the proposition Q is the *consequent*. The language is traditional. Being more informal, P is the "if part" and Q is the "then part." The antecedent is also called the *hypothesis* and the consequent is sometimes called the *conclusion*. Note that "if P, then Q" is, itself, a proposition and may be true or false.

An example of a true conditional with a false antecedent might be "If the weather is nice tomorrow we will all go to the park." If it turns out that it is raining tomorrow and we still go to the park we don't take the conditional as false, since it didn't assert anything about the case it which the weather was poor. We didn't assert that we will go to the park *only if* it is nice. For a conditional "If P, then Q," the related conditional "If Q, then P" is called the *converse*. The truth of a conditional and its converse are independent of each other.

The commutative law of addition in the Natural Numbers is such a conditional: "If x, and y are Natural Numbers, then $x + y = y + x$". To prove the truth of such a conditional, we rely on the definitions of its terms, but also previously proven theorems as well as the axioms defining the system in which we work. We must show that the *consequent*, Q, of the conditional is never false when the *antecedent*, P, is true. If we can do that, we take the conditional itself to be true. But that doesn't mean that either P or Q is necessarily true. Maybe x or y is some other kind of thing than a Natural Number, a puppy, perhaps, in which case the antecedent is false. That eventuality isn't normally considered, both because it isn't useful in mathematics and because it doesn't invalidate the statement of the theorem. So, in proving theorems in conditional format, it is enough to consider the case, or cases, in which the antecedent is true. But if x and y are, indeed, puppies, the consequent might be just meaningless. But if we define "+" for puppies as "likes" then the relation might well not be symmetric. Fido might "like" Sasha, but Sasha might not return the affection. So, context matters, and the antecedent helps capture that, but usually part of the context in just implicit.

The conditional "If John is a living gargoyle, then John is human" is a true conditional, but only because there are no living gargoyles. John may well be human, of course, or not, but neither possibility contradicts the conditional.

We can use sets to illustrate the situation here. Let X_Q be the set of terms for which Q is true. And let X_P be the set of terms for which P is true. To show that the conditional "If P, then Q" is true we want to show that X_P is a subset of X_Q. Note that we are speaking a bit informally here about sets, since "the set of terms for which P is true" might itself be problematic as noted in the previous section, but for most purposes we won't find ourselves in the problem cases. But another way to think of it is that we want to show that the part of X_P that is not in X_Q is empty. So our gargoyle statement is true since X_P is, itself, empty.

Yet another way to think of the conditional is the idea that if the conditional above is true then Q is somehow a more *general* statement than P.

Note that if a conditional is true, but its antecedent is false, then we know nothing about the consequent. It may be true or false. So, the classical conditional, "All men are mortal" which is a conditional, "If x is a man, then x is mortal," says nothing about the situation in which x is not a man. The ancient Greeks, who likely first formulated this statement, believed in gods that were not mortal. Zeus is not a man, so we don't know anything from the statement "All men are mortal" about the mortality of Zeus.

Suppose, however, that we have two conditionals which we assume to be true: "If P, then Q", and "If Q, then P". If both of these are true and if either of P or Q is true then so is the other. Moreover, if either is false, then so is the other. This also says that the sets representing P and Q are the same set. This idea is often expressed as "P if and only if Q." We have seen this above and will again later. "If P, then Q" is the same as "P only if Q." And "P if Q" is the same as "If Q, then P."

A conditional has an alternate formulation. We can restate "If P, then Q" as "If not Q, then not P". Note, that in the set based form this says that if something outside set Q then it must also be outside set P. This is the foundation of proof by contradiction. If we want to show that "If P, then Q" is true, start out by assuming that Q is false and then show that P must also be false. This is how we proved Theorem 3.2. To show that the Natural Numbers is an infinite set ("If X is the set Natural Numbers, then X is infinite."), we first assumed that the consequent was false; assumed X was finite. We then showed that X could not be (all of) the Natural Numbers. The statement "If not Q, then not P" is called the *contrapositive* of "If P, then Q". A statement and its contrapositive are either both true or both false.

But not all theorems have the form of a conditional. Sometimes we know that a conditional is true, perhaps from an earlier theorem, and want to show that its consequent is true also. We proceed by showing that its antecedent is true. Then the truth of two statements, "P" and "If P, then Q", gives us the truth of the third, "Q". The proofs in Chapter 3 that use induction have this structure even when the statement of the theorem does not. For example, Theorem 3.3 which guarantees a predecessor for every natural number other than 1 used induction. We have a true conditional statement; the statement of the induction axiom. It is true, not because we proved it, but because we take it as an axiom defining the basic structure in which we work: the Natural Numbers. Here the conditional is true. We then, in the proof, showed that the antecedent was also true, namely, if x is a Natural Number with a predecessor then so is its successor, x'. This is, itself a conditional, of course, which is the reason that it is a bit complicated to get your mind around, but once we show this, then the consequent of the induction axiom is true: every Natural Number other than 1 has a predecessor. But note, that in the body of the proof, we just proved a conditional to be true, but that conditional was the antecedent of the induction axiom.

As you read the following chapters it will be instructive if you restate our theorems to see if they are really conditionals. Most of them will be. Then decide whether the proof given is direct, showing that when the antecedent is true so must be the consequent, or indirect, by contradiction, showing that when the consequent is false, so must be the antecedent. Or do we use a more complex proof form?

But in all cases, the tools that you may use are the axioms and the already proven theorems, which go back, ultimately, to the axioms.

On occasion it is necessary to introduce a new axiom to make progress. But then you also need to show that the new axiom system is consistent. Hmmm.

A Note About Infinity

Infinity is a tricky subject, even for a mathematician. On the one hand we deal with infinity in math all the time. However, many, not all, mathematicians are careful not to assert the actual existence of an infinite number of things. So, rather than say that the Natural Numbers form an infinite set, they might only assert the related notion that for any Natural Number there is a larger one. They see the natural numbers as going on and on, but not necessarily all existing at once. This is a very subtle distinction, of course.

But there is one situation in which all would agree. A proof of a theorem can't rely on making an infinite number of steps from one statement to the next. In future pages this idea will arise again and I'll be careful to mention, there, that a conclusion can be made using only finitely many choices, for example.

There is, however, a way to get around this problem, for those willing to add an additional axiom. This also assumes that you are willing to accept infinity as something "real." Suppose, for example that I form sets of pairs, each consisting of an odd number and the immediately following even number. The sets, {3, 4} and {9, 10} are examples. There are an infinite number of such sets, of course. Now suppose that I want to choose a number from each of these sets. Is it possible? If I start out selecting something from {1, 2} and then something from {3, 4}, etc. I'll never finish. The *Axiom Of Choice* says that we can do this for arbitrary collections. Not all accept this axiom, and we won't need it here in any case. However it does enable some new ways to prove theorems. The axiom of choice has been shown to be consistent with the Peano axioms, by the way. This is an important consideration whenever you want to add a new axiom.

But in the above example, the axiom of choice isn't needed, nor is it needed for any collection of sets of Natural Numbers (or even Integers, discussed later). This is because we can actually find a rule of selection in these cases. We can, here, for example, just choose the smallest member of the set and are guaranteed that such exists. A different, but definite, rule works for sets of Integers. But in the Rational

Numbers, for example, we need something more. We will explore properties of the Integers and Rational Numbers later in the book.

Sheep Sculpture

Tick Marks

5 Zero. A First Extension – The Whole Numbers

People long understood the concept of *nothing*. But it took a while to associate it with a number. Having no sheep was well understood, if not always desired. It wasn't until positional number systems came along that people needed a place holder for "nothing" in a number representation: 3051.

However, another way to look at this is that we can't solve a simple equation like $x + 3 = 3$ in the Natural Numbers. That is, we can't find a non-empty set of Natural Numbers x that will make the equation true. We *can* solve $x + 3 = 7$, of course. So, some equations can be solved in a given numeric system and others cannot. We shall see this situation again and again here.

Definitions and Properties

Let's define a new value we call *zero*, written 0, that is not equal to any Natural Number. Now form a set that is just the union of two sets: the Natural Numbers and the set consisting of just 0. It is $\mathfrak{N} \cup \{0\}$. We write $\mathfrak{W}$ (W in a Germanic Font) as a symbol for the Whole Numbers: $\mathfrak{W} = \mathfrak{N} \cup \{0\}$. Since 0 is not in $\mathfrak{N}$ we can conclude that $\mathfrak{N} \neq \mathfrak{W}$.

We say that we *embed* the Natural Numbers into a new set, the Whole Numbers. Here the embedding is just the "element of" relationship. Later we will see more sophisticated embeddings, but embedding is an important idea. The subset of the Whole Numbers that are Natural Numbers will behave just as before, but we need to define how the additional element, 0, behaves. In particular, how it interacts with the Natural Numbers. We also need to explore the consequences of our definitions.

We will also extend the successor relation and addition so that it includes zero. In this new set we define the properties of 0 as follows.

$0 + 0 = 0$

$0' = 0 + 1 = 1$

Theorem 5.1: $0 + x = x$ for any Whole Number x.

Proof: For $x = 0$ it is just part of the definition. If x is a Natural Number we use induction on x.

(a) The base case $x = 1$ is also part of the definition.

(*b*) Now suppose that for some $m \neq 0$, $0 + m = m$. Consider m'. Now $0 + m' = (0 + m)' = m'$. We are done. QED

Or are we? Think about it for a moment. Don't just use the facts that you have learned through study of numbers elsewhere. Think about what we really know here, within the axiom system.

Does $0 + x$ even have a definition? How can we say that +, which was defined for Natural Numbers, really makes true the statement $0 + m' = (0 + m)'$?

Actually, we should add this last equality to the definition, in essence defining + in the Whole Numbers by using induction. We extend the definition of + in the Whole Numbers to be consistent with that in the Natural Numbers. Here is a better definition, but note that it defines both the successor relationship and the addition operation:

$0' = 1$

$0 + 0 = 0$

$0 + 1 = 1 + 0 = 1$

$0 + m' = (0 + m)'$ for any Natural Number m.

$m' + 0 = (m + 0)'$ for any Natural Number m.

$m + 1 = m'$ if m is a Natural Number.

$m + n' = (m + n)'$ if m and n are Natural Numbers.

Induction is used to show that the fourth and fifth formulas define $0 + x$ and $x + 0$ for any Natural Number x. Now, the proof of the theorem above is actually sound. The Whole Number 0 is called the *additive identity,* since adding it to any whole number yields the identical number. Similarly $x + 0 = x$ for any Whole Number x. The last two formulas, of course are just the definition of addition in the Natural Numbers.

Exercise: Show that the last four statements in the axioms above remain true if we replace *Natural Number* by *Whole Number*. In other words, show that if m and/or n is zero that the last four statements are true.

Exercise: Prove that if m is a Whole Number then $m + 0 = m$.

Note that this definition required a fairly large number of cases. This isn't considered optimal by mathematicians. It isn't *elegant*. But it is a consequence of defining the Whole Numbers as a simple union of two sets. We need to consider the interactions between the "two parts" of $\mathfrak{W}$.

Exercise: Show that the injection axiom applies to the Whole Numbers. That is, if x and y are in $\mathfrak{W}$ then $x' = y'$ implies that $x = y$. Hint: Use the fact that $\mathfrak{W} = \mathfrak{N} \cup \{0\}$.

Note that we did not just use the above theorem as an axiom. That would have been a statement that $0 + x = x$ for all whole numbers x. That is a statement about an infinite number of things. Instead we give an axiom by which the sum of 0 and any whole number can be constructed from what we know about *smaller* numbers. We shall define *smaller* soon.

Can we be sure anymore that the commutative and associative rules hold?

Theorem 5.2: Addition in the Whole Numbers is Commutative and Associative.

Proof: For the commutative law, we must show that if x and y are whole numbers then $x + y = y + x$. If x or y is zero the previous theorem and the above exercise establishes this. If neither is, then it is the sum of Natural Numbers in which we have the commutative law.

Associativity is a bit easier since $x + y = 0 + (x + y)$, $x + y = (0 + x) + y$ and similarly for 0 in the other positions. So if any of x, y, z are zero the formula reduces to one easy to check and if none are, then associativity in the Natural Numbers gives us the rule: $(x + y) + z = x + (y + z)$.

Theorem 5.3: Extension of Induction. If $P(x)$ is true for 0 and if $P(x')$ is true whenever $P(x)$ is true, then $P(y)$ is true for every Whole Number y.

Proof. This is a consequence of the axiom of induction and the extension of the successor relationship to 0: $0' = 1$. To be more precise: Suppose the conditions of the theorem are met. Then $P(0)$ is true and by the induction hypothesis, $P(0') = P(1)$ is true. But then ordinary induction (the axiom) implies that $P(m)$ is true for every Natural Number. But the Whole Numbers are just 0 along with the Natural Numbers. QED

We shall often use this formulation of induction when discussing the Whole Numbers, rather than giving a separate proof for 0 and 1.

Theorem 5.4: Cancellation Law of Addition: If x, y, and m are Whole Numbers and $x + m = y + m$, then $x = y$.

Proof: By induction on m. If $m = 0$, then $x = x + 0 = y + 0 = y$ by definition of 0. If $m = 1$ this is just the injection axiom. Suppose the statement is true for some k. We consider k'. Assume that $x + k' = y + k'$. Then $(x + k)' = x + k' = y + k' = (y + k)'$. But then, by injection $x + k = y + k$ and so, by the induction hypothesis, $x = y$. QED

You should now look over all of the theorems from Chapter 3. They are still all true about the Natural Numbers. Which of them are true about the Whole Numbers and which are false. How can you modify the statement of each so that the modified theorem is true in the Whole Numbers? Do you need to update any proofs? Note that 1 has no predecessor in the Natural Numbers, but it does in the Whole Numbers.

A new relationship: Less–Equals

Definition: Let x and y be Whole Numbers. We shall say that x is *less than or equal to* y (or, x is *no larger than* y), written $x \leq y$, provided there exists a Whole Number m such that $x + m = y$.

Note that since $x + 0 = x$ it is true that $x \leq x$ for any Whole Number. Because of this we say that the relation $\leq$ is *reflexive*.

Definition 5.5: If $x \leq y$ but $x \neq y$ we say that x is *less than* y and write $x < y$.

Theorem 5.6: If x and y are whole numbers then $x < y$ if and only if there is a Natural Number m such that $x + m = y$.

Proof: If $x < y$ then $x \leq y$ so there is a Whole Number m such that $x + m = y$. But $m \neq 0$ or x would equal y. On the other hand, if there is a Natural Number m such that $x + m = y$, then $x \leq y$ since any Natural Number is a Whole Number. But $x \neq y$ by Theorem 3.8. QED

Note that by this definition $x < x'$ for every Whole Number x, since $x + 1 = x'$.

Also note that $0 < x$ for every Natural Number x, since $0 + x = x$. Likewise $0 \leq z$ for every Whole Number z.

We would like to derive some properties that are implied by these definitions. We shall also define $>$ and $\geq$ as follows:

Definition 5.7: We define $x > y$ to mean $y < x$. This is read as x is greater than y. Likewise $x \geq y$ means $y \leq x$, and is read as x us greater than or equal to y or, x is no smaller than y.

Theorem 5.8: The relation $<$ is *transitive* in $\mathfrak{W}$: If $x < y$ and $y < z$, then $x < z$.

Proof: If $x + m = y$ and $y + n = z$ for Natural Numbers m and n, then $x + (m + n) = (x + m) + n = y + n = z$. Of course $m + n$ is a Natural Number if m and n are. QED

Corollary 5.9: The relation $\leq$ is transitive in $\mathfrak{W}$.

The word *corollary* means a theorem that can be easily derived from some theorem or its proof.

Note that, applying Theorem 3.8, if $x = y$ then it cannot be true that $x < y$.

Also, if $x < y$ then it cannot be true that $x = y$. This is just rephrasing of the above.

Theorem 5.10: If $x, y \in \mathfrak{W}$ and $x < y$ then it is not true that $x > y$.

Proof: by contradiction. If $x < y$ then for some Natural Number m, $x + m = y$. But if it is also true that $x > y$ then $y < x$ and so, for some Natural Number n, $y + n = x$. But then $y = x + m = (y + n) + m = y + (n + m)$, which we have shown impossible (Theorem 3.8) since $n + m$ is a Natural Number. QED

Actually, we can say a bit more than the above. While we have shown that no two of the relationships, $x = y$, $x < y$, and $x > y$, can hold, it is also true that one of them must. This is the Law of Trichotomy.

Theorem 5.11: Law of Trichotomy. If x, y are Whole Numbers then exactly one of the following is true: $x < y$, $x = y$, or $x > y$.

Proof: Let x be an arbitrary Whole Number and let $P(m)$ be the statement that exactly one of the following is true: $m < x$, $m = x$, or $m > x$. We use induction on m.

(a) Case $m = 0$. If $x = 0$ they are equal. Otherwise $m < x$.

(*b*) Suppose $P(m)$ is true then we consider cases for $P(m')$.

(I) If $m = x$ then $m' > x$.

(II) If $m > x$ (and $x + n = m$ for a Natural Number n) then $m' > x$ since $x + n' = m'$

(III) If $m < x$ (and $n + m = x$ for a Natural Number n), then either $n = 1$ or not. If $n = 1$ then $m' = x$. Otherwise the predecessor of n added to m' gives x so $m' < x$. So $P(m)$ is true for every Whole Number m. This completes the proof. QED

Theorem 5.12: The $\leq$ relation is *antisymmetric*: If $a \leq b$ and $b \leq a$ then $a = b$.

Proof: By the law of trichotomy, $a < b$, $a = b$, or $a > b$. If $a < b$ then there is a Natural Number m such that $a + m = b$. But then it is impossible that $a \geq b$. If it were possible, then there is a Whole Number n such that $b + n = a$. But then $b + n + m = a + m = b$ for a Natural Number $n + m$. Theorem 3.8 rules this out. We reach a similar conclusion if we assume $a > b$. QED

Theorem 5.13: The $\leq$ relation in the Whole Numbers is *total*: For any Whole Numbers x and y, either $x \leq y$ or $y \leq x$.

Proof. This is a consequence of the law of trichotomy. QED

A relation like $\leq$ that satisfies the transitive, antisymmetric, and totality rules is called a *total order*. Therefore, the Whole Numbers are a totally ordered set using $\leq$. A relation that satisfies the reflexive, transitive, and antisymmetric laws in a given set is called a *partial order* and the set on which it is defined is called a *partially ordered set* (or *poset*). A totally ordered set is also partially ordered as you can easily show. Note that the set of subsets of a given set A is a poset using $\subseteq$.

Theorem 5.14: If x, y, and z are whole numbers and $x \le y$, then $x + z \le y + z$.

Proof. Since $x \le y$ there is a whole number m such that $x + m = y$. But then $x + m + z = y + z$, so, using the commutative and associative laws, $x + z \le y + z$. QED

Theorem 5.15: If x, y, and z are whole numbers and $x < y$, then $x + z < y + z$.

Proof: As above, except that m is a Natural Number. QED

Exercise: Show that if $x + y = 0$, for Whole Numbers x and y, then $x = 0$.

Some Terminology

We say that the Whole Numbers are *closed under addition*. This means that the sum of any two whole numbers is again a whole number. The Natural Numbers are also closed under addition. The even numbers are also closed under addition, but the odd numbers are not. The sum of two odd numbers isn't odd. We could define a subtraction operation but it wouldn't be closed, since 5 – 8 would have no meaning in the Whole Numbers. We shall later explore this and it will lead us to another extension.

In order to study mathematics efficiently, you need to remember quite a lot of notations as well as definitions. The way to remember things is to use repetition to reinforce your memory. The end of this book will give a few hints about doing that.

Theorem 5.16: $m < n$ iff $m' < n'$, where m and n are Whole Numbers.

Proof:

(a) only if clause: if $m < n$ then $m' < n'$. Proof: If $m < n$ then there is a Natural Number x such that $m + x = n$. But then $(m + x)' = n'$, which can be seen to be $m' + x = n'$ using the associative law.

(*b*) if clause: if $m' < n'$ then $m < n$. Proof: If $m' < n'$ then there is a Natural Number x such that $m' + x = n'$. But $m' + x = (m + x)'$ and so $(m + x)' = n'$. But the injection axiom implies that then $m + x = n$, so $m < n$. QED

Note. We have made a small error above. The injection axiom is stated for Natural Numbers. We have used it as if it applies to Whole Numbers. Does it? Why?

Here is a theorem of a somewhat different sort, though the proof is standard.

Theorem 5.17: Every non–empty subset A of $\mathfrak{W}$ contains a least (smallest) element.

Proof: Let P(m) be the proposition that m is smaller than every element of A. If this is false for 0, then 0 is the smallest element of A, so we may assume otherwise: i.e. $0 < x$ for every $x \in$ A. If P(m') is true for every m for which P(m) is true, then P(m) is always true (induction), but that cannot be, since A is not–empty. So, there is an m such that P(m) is true and P(m') is false. That is to say $m < x$ for every x in A, but $m' \geq x$ for some x in A. Then m' is the least element of A, since there are no Whole Numbers strictly between m and m'. QED

Exercise: Prove the last statement of the above proof. If m and n are Whole Numbers with $m \leq n \leq m'$, then $m = n$ or $m' = n$.

Definition 5.18: A set T is said to be *bounded above* by b, provided $b \geq x$ for every x in T. Then b is called an *upper bound* of T. Note that it may be an element of T.

Corollary 5.19: Every non–empty subset, T of $\mathfrak{W}$ that is bounded above contains a maximum element.

Proof: Let B be the set of all Natural Numbers that are greater than every element of T. Then the bound b is in B so B isn't empty. Let c be the smallest element of B. The value $c \neq 0$ since T is non-empty. The predecessor of c is the maximum element of T. Otherwise its successor, c, couldn't be the smallest element of B. QED

An alternate proof is presented below after a few preliminaries. We noted earlier that we can't define subtraction in $\mathfrak{W}$ as there is no meaning for 5 – 8. But we can define a limited version.

Definition 5.20: Suppose $x \leq b$. Then there is a Whole Number m such that $x + m = b$. Define $b - x$ to be m. Note that it is only defined when $x \leq b$. Note that according to this definition $x + (b - x) = b$.

Lemma 5.21: Let m, n, and b be Whole Numbers with $m \leq b$ and $n \leq b$. Then $m < n$ iff $b - n < b - m$.

Proof. This would be easy in the Integers which we study later. But here we have only a limited form of subtraction. First suppose $m < n$. Then there is a Natural Number k such that $m + k = n$. But then $(b - n) + k + m = (b - n) + n = b$. But, then, by the definition of $b - m$ we get $b - m = (b - n) + k$, so $b - n < b - m$ by the definition of <.

Likewise, suppose $b - n < b - m$. Then there is a Natural Number k such that $(b - n) + k = (b - m)$. But then, adding $m + n$ to both sides and applying commutative and associative laws and the fact that $(b - n) + n = b$ and $(b - m) + m = b$, we get, $b + m + k = b + n$. If we cancel b's from this we see that $n = m + k$ so that $m < n$. QED

Alternate Proof of Corollary 5.19:

Suppose that we have a non empty set S that is bounded above by a Whole Number b. then $b - x$ is defined for every x in B. Let W = $\{y \mid y = b - x$ for some x in S$\}$. Note that W is a non empty set of Whole Numbers. Therefore W has a smallest element z. But $z = b - x$ for some x in S. This element x is the largest element in S by Lemma 5.21. QED

Note that a mapping like this, from x to $b - x$ is often a useful proof technique. Note also that Theorem 5.19 is important and has many uses.

Let's take a moment to consolidate what we know. We have the Peano Axioms, of course, and definitions of an operator (+) and a few relations, such as = and <. Note two things. First, is that without the operators and relations, the Whole Numbers are pretty boring. The most exciting thing about them is that they form an infinite set. It is the operations and relations that add interest. But the operations and relations aren't specified by axioms. They are introduced in definitions. The definitions are simple, but the axioms show that they have complex consequences. Mathematicians hate being bored. So they define things to make life more interesting, and then explore the implications of those definitions.

Now, the big question. Do the Whole Numbers *exist*? Are they real? No, not really. Remember that the universe is finite, so far as is known. The Natural Numbers are a mental construct defined by the human mind. The concepts of *three* and *four* of course, were around before the Peano Axioms, but even those are just mental constructs. How many sheep do you have, Charlie? Ummm. Thirty–five. What is thirty–fiveness, exactly? Well, in the Peano Axiom system it is just the successor of thirty–fourness. But in the real world, it just sheep, or coins, or, ... You could have more or less than thirty–five. But if I have thirty–five coins (with enough value each) and you have thirty–five sheep, then we might be able to make a deal. It is about *matching*. We will explore this later.

Ah yes, coins again. Let's travel back once again to the proposition that any set of coins dropped on a table must land the same way. You have a counterexample, of course, but also want to show why the "proof" is in error. One fruitful way to proceed there is to ask "what is the smallest possible counterexample?" Play with that and see if you can discover the flaw in the *reasoning*. The flaw in the *result* is obvious, of course.

Before we move on, though, let's spend some time and think about what thirty–fiveness, really means. Of course it has to do with the successor relationship, but beyond that, what do we really mean when we say that we have "five" fingers on our left hand. One of the best definitions of the meaning of this has to do with matching. We can match the fingers on our left hand with those on our right. But we can also match them with certain sets of coins, but not others. And with certain sets of sheep but not others. So, in this view, the number five is an abstract property shared between certain sets and not shared with others. *Five* is the *size* property shared with all those sets that can be so matched.

Multiplication

We can define a new operation in the Whole Numbers, multiplication, denoted by a dot ($\cdot$) as follows

Definition 5.22: The *product* of two whole numbers x and y, denoted $x \cdot y$, is defined as follows:

(*a*) $x \cdot 0 = 0 \cdot x = 0$ whenever x is a Whole Number.

(*b*) $x \cdot y' = (x \cdot y) + x$.

Note that we have defined multiplication inductively using things we already know, successors and addition. To "perform multiplication" is "to multiply". The result is called the "product". The two values being multiplied are "multiplicands".

However, we need to be a bit careful about consistency. We have two definitions in the case that we multiply 0 by itself. But they give the same value, so we are fine. Also, we don't need to worry about the first definition conflicting with the second, since 0 isn't a successor, so can't be y' in the second definition.

The fact that $x \cdot 1 = x$ is a simple consequence of the definition (why?). However the following requires proof by induction.

Lemma 5.23: For any whole number x, $1 \cdot x = x$.

Proof: Exercise. Use induction.

Theorem 5.24: Multiplication is Right Distributive over Addition: if x, y, and z are all in $\mathfrak{W}$, then $(x + y) \cdot z = (x \cdot z) + (y \cdot z)$

Proof: By induction on z.

(a) If $z = 0$, then both sides evaluate to 0.

(*b*) Assume as the induction hypothesis, that for some Whole Number m, $(x + y) \cdot m = (x \cdot m) + (y \cdot m)$ and consider m'.

The left side is $(x + y) \cdot m' = (x + y) \cdot m + (x + y)$ by the definition.

By induction, this becomes $(x \cdot m) + (y \cdot m) + x + y = ((x \cdot m) + x) + ((y \cdot m) + y)$ by properties of addition. But, using the definition of multiplication twice, this last is just $(x \cdot m') + (y \cdot m')$ as required. QED

As a convention, we declare that multiplication takes *precedence* over addition. This means that $a + b \cdot c$ is interpreted to be the same as $a + (b \cdot c)$, with the multiplication being performed before the addition. This simply permits expressions to be written more compactly without the parentheses. Similarly, when there is no ambiguity, the $\cdot$ is sometimes omitted in a multiplicative expression. So ab can mean $a \cdot b$: juxtaposition means multiplication. See the associative law below.

Theorem 5.25: Similarly, Multiplication is Left Distributive over Addition:

$x \cdot (y + z) = x \cdot y + x \cdot z$, for any x, y, z in $\mathfrak{W}$.

Proof: exercise. Hint: use induction on x, the fact that $x' = x + 1$, and Lemma 5.23.

Theorem 5.24 above is actually called the Right Distributive Law and theorem 5.25 is the Left Distributive Law. Together they make up the *Distributive Law*. The Theorem 5.27 (below) seems like it would be easy, since its statement is simpler than the above. The proof, however depends on the Distributive Law.

Theorem 5.26: Multiplicative Identity: If x is a whole number, then $x \cdot 1 = 1 \cdot x = x$.

Proof: By the definition of multiplication and the fact that $0' = 1$: $x \cdot 1 = x \cdot 0' = (x \cdot 0) + x = x$. Lemma 5.23 gives the other result. QED

Since $1 \cdot a = a \cdot 1 = a$, 1 is called the *multiplicative identity* for the Whole Numbers, just as 0 is the *additive identity*. The number *one* is also often called the *unit element*.

Theorem 5.27: Multiplication is Commutative: $x \cdot y = y \cdot x$ for Whole Numbers x and y.

Proof: By induction on y in $x \cdot y$.

(a) If $y = 0$, this is part of the definition of multiplication.

(*b*) Assume $x \cdot y = y \cdot x$ and consider y'. Then $x \cdot y' = x \cdot (y + 1) = x \cdot y + x = y \cdot x + x = x + y \cdot x = 1 \cdot x + y \cdot x = (1 + y) \cdot x = (y + 1) \cdot x = y' \cdot x$. QED

Exercise: List all of the theorems we used in this proof.

Theorem 5.28: Multiplication is Associative: If a, b, c are in $\mathfrak{W}$, then $a \cdot (b \cdot c) = (a \cdot b) \cdot c$

Proof: Induction on c.

(a) If $c = 0$, both sides are 0.

(*b*) Assume $a \cdot (b \cdot c) = (a \cdot b) \cdot c$ and consider c'.

$a \cdot (b \cdot c') = a \cdot ((b \cdot c) + b) = a \cdot (b \cdot c) + a \cdot b = (a \cdot b) \cdot c + (a \cdot b) = (a \cdot b) \cdot c'$. QED

If a whole number x is the product of other whole numbers those numbers are called the *factors* of x. The number *one* is always a factor of course as is x itself.

These are *trivial factors*. Some numbers have non–trivial factors. For example $6 = 2 \cdot 3$. Whole Numbers with nontrivial factors are called *composite*. Other numbers have only trivial factors, such as 17. Such numbers are called *prime*. It is known that there are an infinite number of primes, hence there is no largest prime, but it takes more than what we have here now to prove it. The proof is by contradiction: Assume there are only finitely many primes and then use that fact to show there must be another that can't be equal to any of them. We will return to this later.

Definition 5.29: A Whole Number that has x as a factor is called a *multiple* of x.

Note that in the Whole Number domain, there are simple equations that have no solution at all. One of them is $x + 5 = 2$. Can you prove that this has no solution? Use the theorems above and induction if necessary.

Likewise, we can write inequalities. One such as $x < 5$ has several solutions. The equation $x > 5$ has infinitely many. As a shorthand we often say something like $5 < x < 10$ to mean $x > 5$ *and* $x < 10$. There are four solutions, of course in the Whole Numbers. The inequality $x < 0$ has no solution in the Whole Numbers.

In algebra there are several interesting manipulations you can do. For example if $x = y$ and you add the same value to both x and y the results are also equal. This is because sums are unique. Thus, $x + m = y + m$ if $x = y$. A similar law holds for multiplication. If $x = y$ then $x \cdot m = y \cdot m$. Can you provide a proof? A consequence of this is that you can substitute one value for another, if they are equal, in sums and products. This is the Law of Substitution.

Is it true that if $x \cdot m = y \cdot m$ then $x = y$? When is this true and when is it false?

Well, if $m = 0$, it can't be true since both sides are zero independent of the values of x and y. So we need to exclude that case.

Theorem 5.30: Cancellation Law of Multiplication. If x, y and m are Whole Numbers with $m \neq 0$, and $x{\cdot}m = y{\cdot}m$, then $x = y$.

Proof: By induction on m, starting with $m = 1$. If $m = 1$ then $x = y$ by the Multiplicative Identity Theorem (Theorem 5.26). Assume the theorem is true for a value k and consider k'. Suppose $x \cdot k' = y \cdot k'$. Then $(x \cdot k) + k = (y \cdot k) + k$. By the cancellation law for addition, $x \cdot k = y \cdot k$ and the induction hypothesis gives $x = y$. QED

These cancellation rules for addition and multiplication are tools that help in solving equations. Suppose we have the equation $6 \cdot x + 7 = 31$. Re–writing it as $6 \cdot x + 7 = 24 + 7$ we see that $6 \cdot x = 24$. But this is $6 \cdot x = 6 \cdot 4$, and so $x = 4$ is the (only) solution. Note that we have not had to introduce subtraction or division to do this.

Exercise: State and prove the *substitution law of multiplication*. Note that the substitution and cancelation laws of multiplication are not converses of each other. You can substitute zero, but not cancel it.

With substitution laws for both addition and multiplication you can substitute an equal value into a given expression quite freely. You may need to use induction in the proof. But in the extensions to the Whole Numbers that we see later in the book, the proofs of cancellation laws quite easily reduce to these laws for Whole Numbers.

Theorem 5.31: If $x > 1$ and y are Natural Numbers then $x \cdot y > y$.

Proof. By induction on y. If $y = 1$ then $x \cdot y = x > 1 = y$. The base case is proved.
Assume the theorem is true for k. We consider k'. Now $x \cdot k' = x \cdot k + x$.

But $x \cdot k > k$ so $x \cdot k' = x \cdot k + x > k + x > k + 1 = k'$. QED

What is Easy, What is Hard, and Mathematical Insight

So far we have been going on theorem, prove, theorem, prove, definition, theorem, etc. While it is technically correct, it may not be especially satisfying. Many courses in mathematics are like that, however. But a mathematician needs more: insight into why things might be the way they are. In order to *create* mathematics, you need this insight. "What might be true (is probably true) that I don't *know* is true?" This is what drives mathematics forward. When Peano came up with the axioms we have studied, the world had no doubt about what the facts of the Natural Numbers are (though the art has been advanced since then). It was just a new way of looking at the facts as a logical system.

Note that the definition of multiplication, done inductively, used induction on the right factor only. We could have done it on the left factor equivalently, but not on both. That would have left us open to the possibility of introducing inconsistencies. But since we defined it from one side only, in essence, it made it difficult to talk about the commutative law since the factors need to be reversed there. But with the insight that multiplication is really just repeated addition and we have a commutative law for addition, we might be able to use that to help. But before we could do that, we need the distributive law, which relates the two operations, but doesn't require reversing any operands. (The inputs to an operation are called *operands*.)

So, a mathematician has certain tools. These consist of axioms, proven facts, definitions, proof mechanisms, etc. but the most important is insight. The same is true of any craftsman or craftswoman. A carpenter has a box of tools. Which tool is correct at this moment? The big hammer or the small? Skill and experience will tell you which one to use. But the deeper question is "What *should* be built?". This is even more critical for the artist, whose goal is more internalized. Think of mathematics as an art form. Artists, like mathematicians, gain skill from hard work; learning to use the tools; learning when a given tool should be applied; studying the work of other artists, and so on. But, what should I build?

Notice also that when you define something new in a mathematical system you then need to consider its interactions with the other things already defined. So, addition is commutative and associative. Then you add multiplication, also commutative and associative. But then, addition and multiplication interact in some way if the system is at all interesting. Ah ha. Multiplication distributes over addition. Does addition distribute over multiplication as well. .... No, it doesn't. But in set theory, it is interesting that set union distributes over intersection, but intersection also distributes over union. Very interesting.

There are two different sorts of properties of mathematical entities: those that are defined and those that are derived. In Peano's axioms we defined a few properties and then derived other properties from them. Successor is defined. The infinite size of the set of Natural Numbers is derived. We want to keep the number of defined properties to be small and simple. Otherwise we increase the chance that the definitions will be inconsistent. Suppose we define too much and discover one derivation of 2 + 2 that gives 4, but a different one that gives 5. Something is wrong, but what. Minimizing definitions helps to avoid this problem, but it never goes away completely unless you can provide a proof of consistency.

An interesting set of numbers are called *factorials*. They have lots of factors. The factorial of a Whole Number n, written n!, is the product of all of the Whole numbers between 1 and n. The product of the elements of a set. 4! = 24. By convention, 0! = 1. If you have n people, in how many ways can you form a single line (queue) containing everybody. This is a *choice problem*. You have n choices for the first person, n-1 for the next, etc., but only one person left to be last. So there are n! different ways to do it. As n increases n! increases very fast. A somewhat related problem, also solved with factorials is to determine how many subsets of, say, 5 there are in a set of, say, 20 elements. This is a *counting problem* and there is a subfield of Mathematics, called Discrete Mathematics devoted to it. It is important in probability, statistics, and computing.

Algebra, Variables, Unknowns, and How Mathematics is Not Computer Science

We have used symbols like x, y, and m to represent arbitrary numbers in most of the proofs above. If A is a set (there we go again) and we want to discuss either a member that we don't wish to name explicitly, or a representative member, we use such a symbol. Often we call them *variables*, but they don't really *vary* (i.e. change). All that is meant is that, for our purposes at the time, which specific element we choose isn't important. Computer Science uses the word differently, however. A variable in, for example, the Ruby Programming language is a symbol that can be used to identify a *specific* value, but the value might change as the program executes. The value with which the name is associated will vary over time. This is not generally true in mathematics. In mathematics, the value is not specific. In programming it is. In mathematics it doesn't change (within a proof). In computing it changes.

The term Algebra has many meanings. For now we will just consider it to be a way to manipulate numeric quantities symbolically, rather than numerically. It is focused on the operations. We have done this above in writing $x + 1$, for example. The simple things we do in Elementary Algebra have been generalized, however, into a major branch of mathematics. One thing we do in algebra is to write equations and inequalities. For example, in the Whole Number domain, we could write the equation: $x + 3 = 5$. The variable here is x, but there is no notion of it varying. In a context like this it is sometimes called an *unknown*. The meaning of that comes from the fact that when we write such equations they can be true or false and we would like to know when they are true. We seek a value (or several) that is a solution of the equation: one which, when substituted for the variable, makes it true. This is what we mean by solving an equation or inequality. It isn't really unknown except in the sense that until we solve the equation we don't know the answer. Here, the only solution is $x = 2$. That is to say, the substitution of 2 for x makes the equation into $2 + 3 = 5$ which is true. It is a bit more work to show it, but any other substitution makes it false. So, *variable* doesn't mean *varies* and *unknown* only means not known *yet*. It is just a convention that arose over many years.

Some computer languages use = as an operation, rather than a relation. A statement in the *c* programming language like

$x = 5 + y;$

is not an equation with two variables. It is an operation that says the value associated with x should be changed to the value 5 added to the current value of y. It is an operation that changes the value of x, not an equation to be solved. It is good not to confuse the two ideas, even though the same terminology may be used. Mathematics is used in computing and computing sometimes aids in mathematics, but the basic ideas are very different.

Some Interesting Sums

Induction can help us to compute some generalized sums. We want to add up a set of numbers of some particular form. Some of these sums are very important. One that is interesting is the sum of the first k odd numbers. The sum is k^2. So $1 + 3 + 5 + 7 + 9 = 25$. We can prove this in general using induction. Clearly it is true for the case 1. The sum of the first 1 odd numbers is just 1 and this is 1^2. Now suppose that it is true for some k: $\sum\{2n-1 \mid 1 \le n \le k\} = k^2$. Now add the next odd number to both sides. The left becomes $\sum\{2n-1 \mid n \le k + 1\}$ and the right is $k^2 + 2k+1$. But this is just $(k + 1)^2$. One of the first things I ever did in maths is to analyze how to use this to compute square roots using subtraction.

Another sum with more importance, generally, is the following. I'll start from zero here, rather than 1. We will also use quite a lot of things not yet introduced, such as Real Numbers and division, etc., but I want to emphasize how induction can be useful. Let r be a number different from 1. Then $\sum\{r^n \mid 0 \le n \le k\} = (1 - r^{k+1})/(1 - r)$. For example, $2^0 + 2^1 + 2^2 = 7 = (1 - 2^3)/(-1)$. Here / indicates division. The formula is true if $k = 0$. That is, the left side is just $r^0 = 1$. The right is $(1 - r^1)/(1 - r) = 1$, since $r \neq 1$. Now suppose that the formula is true for some k: the sum of the first $k + 1$ powers of a number other than 1 is $(1 - r^{k+1})/(1 - r)$. Yes there are $k + 1$ terms on the left since we started at 0. Now add the next power of r to both sides. The right side is $(1 - r^{k+1})/(1 - r) + r^{k+1}$. Again anticipating rules you already know, but we haven't yet proven, this is the same as $(1 - r^{k+1})/(1 - r) + r^{k+1}(1 - r)/(1 - r) = ((1 - r^{k+1}) + r^{k+1}(1 - r))/(1 - r) = ((1 - r^{k+1}) + r^{k+1} - r^{k+2}))/(1 - r) = (1 - r^{k+2})/(1 - r)$. And this is what we required. And note that if r is a real number between 0 and 1, not just different from 1, that the value of r^{k+1} is very small for large k. We will introduce Real Numbers later. The sum here is called a *geometric sum*, by the way. While the geometric sum formula holds for a wider range of numbers, our proof is only valid for Whole Numbers.

Exercise: $\sum\{n \mid n \in \mathfrak{N}, 1 < n < k\} = k \cdot (k + 1)/2$. In particular, $1 + 2 + 3 + 4 = 4(5)/2 = 10$. Note that either k or k + 1 is even so the result of the division on the right side is always an integer.

Representing Whole Numbers

There are many ways to represent individual numbers. One of the oldest and simplest is to use tick marks. If you see six sheep, you make six marks on a rock ''''''. Of course 6 = 0'''''', so this makes sense to us. But distinguishing 300, from 305 sheep is difficult with tick marks, so people devised other means. But your friend might have counted his flock of sheep by making *x*'s on a piece of bark. When you get together you could compare the sizes of the flocks by matching tick marks and *x*'s. If they don't all match, someone has more sheep.

The Inca people, in the Americas used knotted cords: quipu. The Romans invented a sophisticated system, that is, nevertheless, difficult to calculate with. In China, on the other hand, the abacus makes calculation easy, but at the expense of compact and stable representation. If you drop the abacus you lose your calculation, just as if your calculator runs out of battery power.

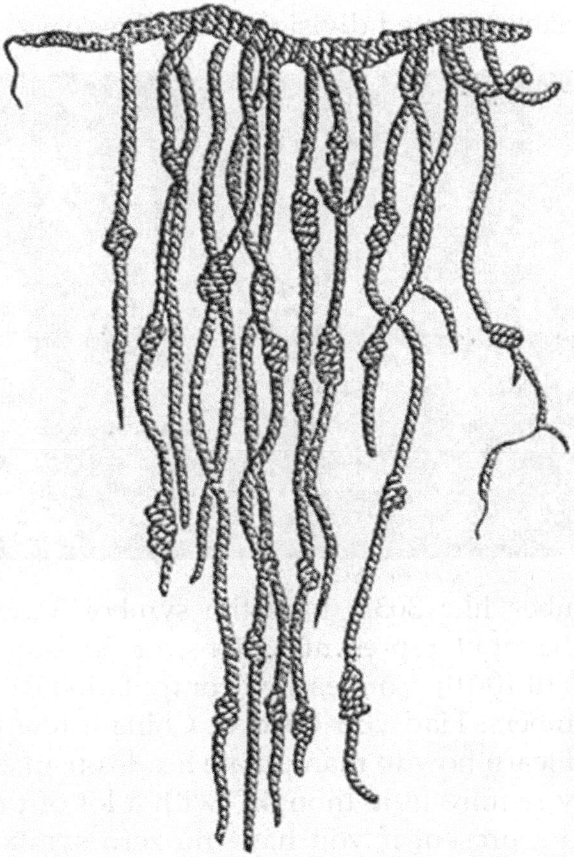

Quipu Example

Modern positional number systems, such as our base ten system are nearly 2000 years old. These are often said to have originated in Arabic lands because they came to Europe via Arabia. In reality, it was originally a Hindu invention, adopted by Arabic and Persian scholars. A positional numbers system with a given base, requires knowledge of zero as well as a symbol to represent it. If the base is n (say ten) it also requires n individual symbols whose value depends on where it appears in a written number. The usual "Arabic Numerals" are 0, 1, … 9, of course. Any base will do (except 0), though base 1 is a special case. It isn't precisely known why we use base ten, but is assumed to be because we have ten fingers. There are advantages, however, to other bases. Base twelve has the advantage that the base itself has a lot of factors. A base that is a power of two, such as two or sixteen, has the advantage that it makes repeated division into halves easy.

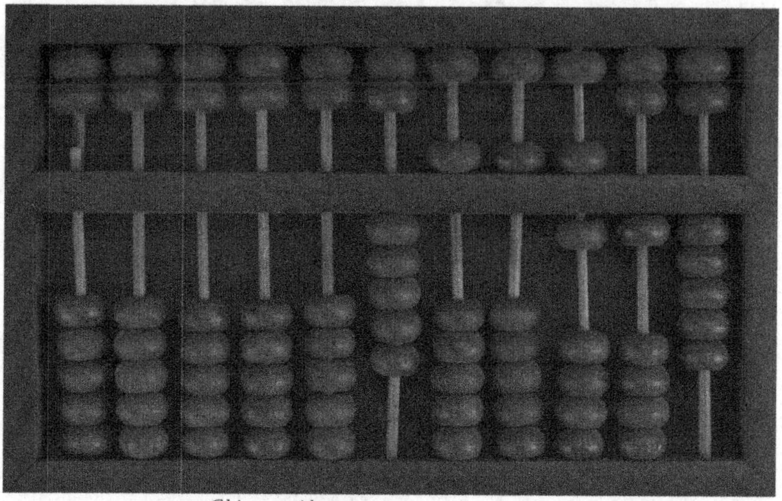

Chinese Abacus

Using base ten, a number like 3037, uses the symbol 3 twice, with different meaning. The one to the right represents 3 tens, or 30, but the one to the left represents 3 thousands (10·10·10). You learned computational algorithms in school to manipulate such numbers. Had you lived in China a few hundred years ago (even today) you would learn how to manipulate beads on an abacus that have the same effect, but actually require little thought (with a lot of practice). Notice that 3037 is very difficult to represent if you have no zero symbol. Initially a blank space was used: 3 37, but this is open to errors of interpretation, especially at the right end. The difference between 3 and 300 is essentially invisible using just space. In Roman Numerals 3037 is MMMXXXVII. V always stands for five. If you mean 500, you write D, instead. The system isn't difficult for reasonably sized numbers, but is not especially computational. It is also limited to numbers less than four million.

In base 5, for example, the symbols 0, 1, 2, 3, and 4 can be used. The symbol 24_5 means 2 fives plus 4 units or 14_{10}. Likewise 124_5 is twenty-five (one five-squared) plus 2 fives plus 4 or thirty-nine. Commonly used bases are two (binary), eight (octal), ten (decimal), and sixteen (hexadecimal). Base sixty has appeared historically. It is called sexagesimal and was used in Sumeria and later in Babylon around five thousand years ago. We take much of our time and angle measurements from that usage.

They say that if you ask a horse to count to 10, by stamping on the ground, you hear "stamp, stamp, stamp, stamp," assuming the horse got it right.

It is worth a bit of study and exploration, but note that the algorithms you learned for base ten computations, also work in other bases in the same way. In adding numbers in base seven, for example, you carry 1, but when you reach seven, just as you carry 1 when you reach ten in base ten. It is all based on the idea of a polynomial, which we shall explore later. But for now, note that a numeric value is the same, no matter how we represent it.

Summary So Far

To define the Whole Numbers we have added a single value to the Natural Numbers. The extension was a bit awkward, actually, though it followed the historical development. It might have been easier to just throw out the Natural Numbers and start with the Whole Numbers and develop the Peano Axioms with the statement that "zero is a Whole Number." Perhaps the awkwardness was instructive, though. Adding things piecemeal can be messy. But we now have a set, $\mathfrak{W}$, two operations, addition and multiplication, and three relations, equals, less-than, and less-equals. Most important we have developed a set of properties of these operations and relations in the various theorems. These properties, such as the commutative and associative laws seem to be fundamental to our understanding of numbers. As we move on we shall want to see how these properties are preserved, and not preserved, in the more complex systems we will develop. We will see similar theorems in later sections that are analogs of what we have in the Whole Numbers. And we shall note when an important theorem in one system does not hold in a similar system. This is one of the things that makes math interesting. But developing the corresponding theorems in larger systems can be tedious, even if necessary. It is likely true that any field of human study has both exciting and boring elements.

Hmmm. A math game found in an ancient cave...

6 The Integers

We would like to extend our number system again in a particular direction. While there are many things that we could do with just the Whole Numbers this particular extension will permit something we found impossible in the previous chapter. We noted that some simple equations like $5 + x = 3$ have no solution in the Whole Number system. Of course, the facts you know already say that the solution should be –2, but this is not a Whole Number by our definition.

There was some notion of negative numbers known to the ancients. More than 2000 years ago the Chinese had a good notion of debits and credits. But they didn't have a distinct way to represent negatives other than by using different colored symbols for debits and credits. It took until about 1300 years ago, or thereabouts for a solid set of ideas to arise in India. I assume the ancients could even subtract, though they were careful to subtract only a smaller number from a larger one. But that is just my own speculation.

The path we take in this development may seem strange at first. We will wait a bit to motivate it. We take this path, strange as it may seem, because it makes a lot of things easier, though that won't be apparent at first. But, just as when we extended the Natural Numbers to the Whole Numbers by embedding the former into the latter, we will embed the Whole Numbers into a more complex structure, while retaining their essential properties.

Consider the set $\mathfrak{W}^2 = \{ (x, y) \mid x \text{ and } y \text{ are Whole Numbers} \}$. This is a set of *ordered pairs* of Whole Numbers, such as (5, 2), (0, 3), etc. It is an infinite set, of course. We will use this as the basis of the Integers, though we need a bit of work first. In the set $\mathfrak{W}^2$ we will need to define a few things. But first note that (5, 2) is not the same ordered pair as (2, 5). The ordering is important, hence the name.

Definition 6.1: Successor. If (x, y) is in $\mathfrak{W}^2$, then $(x, y)' = (x', y)$.

Note that (x', y) is defined since x and y are whole numbers. This defines $(x, y)'$, the successor of (x, y). So, in particular, $(3, 5)' = (4, 5)$.

Definition 6.2: Sum. If (x, y) and (w, z) are in $\mathfrak{W}^2$, then the sum is defined to be $(x, y) + (w, z) = (x + w, y + z)$.

Definition 6.3: Product. If (x, y) and (w, z) are in $\mathfrak{W}^2$, then the product is defined to be $(x, y) \cdot (w, z) = (x \cdot w + y \cdot z,\ x \cdot z + y \cdot w)$.

Well that was easy, but what *ever* could it mean? Note initially, that we aren't treating a specific ordered pair as a number. It is a bit more complex than that.

Exercise: Show a few examples of these definitions.

It is messy, but not difficult to show that + and · in $\mathfrak{W}^2$ satisfy the laws we defined above. + and · are commutative and associative, and · distributes over + in the usual way. I will leave it as an exercise. You don't even need induction. Use the definitions and the properties of $\mathfrak{W}$ to prove the theorems.

The motivation for this seemingly strange definition is as follows. Suppose we are more concerned with accounting than with mathematics and we want to know who is the richer farmer, measured in number of sheep. If John has 5 sheep and Mary has 7, we could represent this as (5, 7). We conclude that Mary has two more sheep than John, so is richer since there is an excess of 2 in the second slot. But we can manipulate these quantities. Suppose we reduce each value in the pair by 1. We get (4, 6), which has the same excess. Suppose we do this repeatedly until one of the values is 0: $(3, 5) \to (2, 4) \to (1, 3) \to (0, 2)$. Now the excess is explicit. In fact, bringing it back to things you know already, these pairs are a way to represent the integer you know as –2. However, we won't use these pairs directly, since there would be an awfully lot of "symbols" to represent any given number.

Another interpretation is debits and credits. Suppose Jamal owes Ahmed 5 drachmas. Ahmed can represent this as (0, 5). When Jamal pays a bit of the debt, say 2 drachmas, Ahmed can change the debt to (2, 5), retaining both what was originally owed and what has been paid.

There is a further, educational, motivation in our approach. In the next section we will explore *equivalence relations* and *equivalence classes*. These have proven to be very fruitful in advancing mathematics, so every serious student should know something of them. They are a bit abstract, also, so they let us stretch our abstraction muscles.

Look at the definitions of the operations above and consider what happens if we apply them only to pairs that have 0 in the *second* slot. $(x, 0)' = (x', 0)$, $(x, 0) + (w, 0) = (x + w, 0)$, and $(x, 0) \cdot (w, 0) = (x \cdot w, 0)$. This is consistent with the definitions in the Whole Numbers. Explore a bit more using only pairs of the form $(0, x)$.

Before we define the actual *Integer* domain, we need to take a side trip.

Equivalence Relations and Equivalence Classes

Back when we first discussed equality in the Natural Numbers we defined it as a relationship between Natural Numbers that is reflexive, symmetric, and transitive. We wish to generalize that now. Any relation that satisfies these three rules is called an *Equivalence Relation*:

Definition 6.4: A relation $\sim$ between values of a set A (i.e. in A^2) is called an *Equivalence Relation* provided

(reflexive) $x \sim x$ for all x in A,

(symmetric) If x and y are in A and $x \sim y$, then $y \sim x$,

(transitive) If x, y, and z are in A and $x \sim y$ and $y \sim z$, then $x \sim z$.

Note that we will often use $\sim$ as a generic name for a relation.

Definition 6.5: Let $\sim$ be an equivalence relation on a set A. Let x be an element of A. Then the *equivalence class of* x, $E(x)$, is $\{ a \mid a \in A \text{ and } x \sim a\}$.

The only element of the equivalence class of x using the equality relation in $\mathfrak{W}$ is x itself. Only x is equal to x. But that isn't necessarily the case. The equivalence classes of some equivalence relations are quite large as we shall see.

Equivalence classes have some interesting properties. One is that they are not empty since a, itself, is in its own equivalence class, $E(a)$, by the reflexive rule.

We noted above that equality between elements of $\mathfrak{W}$ is an equivalence relation. In contrast, < is not an equivalence relation. It isn't symmetric. Neither is ≤. The latter relation is reflexive and transitive, though. So, in some ways these properties are more important than the relations themselves, since in different combinations they give us different, hopefully interesting, things to study.

We haven't emphasized it in our notation, but E(x) is relative to a specific equivalence relation. Should it become necessary to be more definite, we might write $E_{\sim}(x)$ for an equivalence class associated with relation ~.

Theorem 6.6. Let ~ be an equivalence relation on A and let x and y be in A. Then either E(x) = E(y) or the two classes are disjoint (have no common elements.)

Proof: Suppose the two sets are not disjoint. Then there is a z in both sets. Let t be any member of E(x). Then $t \sim x$, and $x \sim z$, so $t \sim z$. But z is also in E(y), meaning z ~ y, so $t \sim y$ and t is in E(y). Thus E(x) is a subset of E(y). Similar reasoning shows E(y) is a subset of E(x), so the two sets are identical. QED

The equivalence class of x is all those items equivalent to x. If value b is not equivalent to value x then its own equivalence class, which many have many elements, is completely disjoint from that of x. An equivalence class is then like a distinct object that we can work with. We shall treat it as a value.

A note on "proof": Many of the proofs given below are outlines only. Details are purposely left missing. You may be able to mentally fill in the details yourself, or you may need to spend some time to work through them.

Definitions and Properties

Let's define a particular relation on $\mathfrak{N}^2$ and then show it is an equivalence relation. We will use α to denote the relation, but it is not a standard name, unlike = which is, of course, a standard name for a well-known relation.

Definition 6.7: Let (a, b) and (c, d) be in $\mathfrak{N}^2$. Define (a, b) α (c, d) provided $a + d = b + c$. You can just read (a, b) α (c, d) as "(a, b) is related to (c, d)".

Theorem 6.8: The relation α is an equivalence relation in $\mathfrak{N}^2$.

Proof: Let a, b, c, d, e, f be in $\mathfrak{N}$.

(reflexive) (a, b) α (a, b) since $a + b = b +$ a.

(symmetric) If $(a, b)\ \alpha\ (c, d)$ then $a + d = b + c$, but this is exactly what is required for $(c, d)\ \alpha\ (a, b)$ by the commutative law in the Whole Numbers.

(transitive) If $(a, b)\ \alpha\ (c, d)$ then $a + d = b + c$. If $(c, d)\ \alpha\ (e, f)$ then $c + f = d + e$. Adding f to both sides of the first equality and b to both sides of the second, we see that $a + d + f = b + d + e$. By the cancellation law of addition in $\mathfrak{W}$, we have a $+ f = b + e$, so $(a, b)\ \alpha\ (e, f)$. QED

Definition 6.9: The equivalence class of (x, y) in α will, in this chapter, be denoted $E(x, y)$, or $E_\alpha(x, y)$ if we need to be more precise. Note that it is more technically correct to write $E_\alpha((x, y))$ since it is the equivalence class of an *ordered pair* of elements of $\mathfrak{W}$,

Each equivalence class represents what we shall call an Integer. The integer 1 is that class containing (1, 0), for example, though it also contains (5, 4), (10, 9) and infinitely many others.

Definition 6.10: The set of Integers, $\mathbb{Z}$, is the set of equivalence classes defined by α on the set $\mathfrak{W}^2$. The letter Z comes from the German: Zahl, number.

Let's say it another way. The equivalence class of (1, 0), E(1, 0), in the set $\mathfrak{W}^2$, is a thing we can name. We name it 1, though that name has been used for other things, of course. Likewise 2 is now a name for the equivalence class containing (2, 0), etc. We call the equivalence class containing (0, 2) by the name –2, and, in general, the equivalence class of $(0, n)$ by the name $-n$, for n in $\mathfrak{W}$. Thus the equivalence class of (0, 0) might be called either 0 or –0 and these are the same value. Note that the symbol "-" used here is not yet related in any way to an operation. It is just a mark to indicate that 2 is not the same as -2.

We can use the definitions of operations that we gave above for pairs to induce equivalent operations on the Integers.

Definition 6.11: Let $E = E(a, b)$ be some equivalence class (relative to α) in $\mathbb{Z}$. Then $E' = E((a, b)')$. In words: the successor of an equivalence class is the equivalence class of the successor.

We need to show that this has meaning. What if we start with two different representatives of E. Do we get the same result? If not the above is not a consistent definition.

Theorem 6.12: Let (a, b) and (c, d) be two representatives of the same equivalence class (i.e. they are equivalent to each other). Then $(a, b)' \alpha (c, d)'$.

Proof: $(a, b)' = (a', b)$ and $(c, d)' = (c', d)$. But if $a + d = b + c$, then $a' + d = b + c'$ using properties of Whole Numbers. Thus $(a, b)' \alpha (c, d)'$. So the successor operation gives the same result for any element of an equivalence class, since $(a, b)'$ and $(c, d)'$ are in the same equivalence class. QED

Definition 6.13: Let E = E(a, b) and F = E(e, f) be two equivalence classes in $\mathbb{Z}$. Define E + F = E((a, b) + (e, f)). In words: the sum of two equivalence classes is the equivalence class of the sum.

Again we need to show that this definition is consistent; that it gives the same result if we choose different representatives of E and F.

Theorem 6.14: Let $(a, b) \alpha (c, d)$ and $(e, f) \alpha (g, h)$. Then $((a, b) + (e, f)) \alpha ((c, d) + (g, h))$.

Proof: Exercise.

Definition 6.15: Let E = E(a, b) and F = E(e, f) be two equivalence classes in $\mathbb{Z}$. Define E · F = E($(a, b) \cdot (e, f)$). Or: the product of two equivalence classes is the equivalence class of the product.

Again, a consistency check is required.

Theorem 6.16: Let $(a, b) \alpha (c, d)$ and $(e, f) \alpha (g, h)$. Then $((a, b) \cdot (e, f)) \alpha ((c, d) \cdot (g, h))$.

Proof: Exercise.

These theorems justify our use of the equivalence classes as if they were values. To operate on a value choose any representative element of its class, then perform your operation and use the equivalence class of the result. It will give the same result had you chosen a representative of the class differently.

Exercise: Let x = E(a, b) be any Integer. Show that E(b, a) + E(a, b) = E(0, 0). Note that if E(a, b) is an equivalence class, then E(b, a) will be denoted by -E(a, b) = $-x$ and is called the *additive inverse* of E(a, b). So, E(3, 5) = -E(5, 3).

The terminology used by mathematicians for the situation described above, is that the three operations are *well-defined*. We need these theorems, since to compute a sum in the Integers we have many different ways to proceed, since we can take *any* representative of an equivalence class. But since they all yield equivalent results, the operations always give a single result.

Theorem 6.17: Addition is commutative in $\mathbb{Z}$.

Theorem 6.18: Addition is associative in $\mathbb{Z}$.

Theorem 6.19: Multiplication is commutative in $\mathbb{Z}$.

Theorem 6.20: Multiplication is associative in $\mathbb{Z}$.

Theorem 6.21: Multiplication distributes over addition in $\mathbb{Z}$.

Proofs: Exercise. Look at representative pairs and use the properties of Whole Numbers.

Theorem 6.22: Additive Identity: If x is an Integer then $x + 0 = 0 + x = x$.

Proof: 0 = E(0, 0). The rest is easy.

Note that 0 is represented by any pair with equal components: (n, n).

By the exercise above, every Integer x has an additive inverse, $-x$.

Theorem 6.23: Multiplicative Identity: If x is an Integer then $x \cdot 1 = 1 \cdot x = x$.

Proof: 1 = E(1, 0). The rest is easy.

We mentioned above that (1, 0)'s class is the Integer 1. It behaves just like the Natural Number 1 with the above operations. For a number like (7, 3) you can reduce it until the second component is 0, to see that this is like 4. But what about (3, 5)? If we reduce this, then we get (0, 2), which behaves like –2. In fact we define the symbol –2 to be the equivalence class of any pair in which the second component is two more than the first.

Note that (5, 7) + (8, 6) = (13, 13) α (0, 0). So, (–2) + 2 = 0 as you expect.

Why did we go to all of this trouble just to define negatives? There are two reasons, really. One is that I wanted to introduce equivalence classes and isomorphism (see below) since both have many other uses. The other is that, once we establish that the Whole Numbers really are a subset of the Integers then the proofs of the remaining theorems will be easier, since we don't have to make special cases for the whole numbers and the other (negative) numbers. A naïve approach, using a simple set union makes the theorem proofs very messy. Recall that in extending $\mathfrak{N}$ to $\mathfrak{W}$ we did require special cases to handle 0. But here there would be many more special cases. We avoid that with the equivalence class construction.

Definition 6.24: If E(a, b) and E(c, d) are in $\mathfrak{Z}$, then E(a, b) ≤ E(c, d) provided there is a whole number k such that E(a, b) + E(k, 0) = E(c, d). We say that x = E(a, b) is *less than or equal to* y = E(c, d).

Likewise, we say that one Integer x is less than another, y, $x < y$, provided $x \leq y$ and $x \neq y$.

Thus –5 < –3, since –5 + 2 = E(0, 5) + E(2, 0) = E(2, 5) = E(0, 3) = –3.

Exercise: Show that 5 < 7 and that – 3 < 0.

Exercise: Show that E(0, a) < E(b, 0) for any Natural Numbers a and b.

Exercise: Show that $x < y$, for x, y in $\mathfrak{Z}$, if and only if there is a natural number w such that x + E(w, 0) = y.

Exercise: (Multiplication is monotonic): Show that if x, y, and z are Integers with x < y and z > 0, then xz < yz.

One problem we are left with is the fact that we haven't really *extended* the Whole Numbers. We have created an entirely new kind of thing. We need to show that the Integers contain a subset that is *just like* the Whole Numbers. But *just like* requires a definition and the *showing* is a proof.

Definition 6.25: An *isomorphism* is a structure preserving bijection between two sets. If there is an isomorphism between two sets we call them isomorphic, or isomorphic to each other.

Well, that *is* the abstract definition, but it will require a bit of explanation, extending our detour. We will return to the discussion of Integers.

Definition 6.26: A function from A to B is called *structure preserving* if some set of operations of interest in A is mapped into corresponding operations in B in such a way that they are indistinguishable in the two sets.

Well, that isn't really a definition, but an idea. It depends a lot on some indeterminate words like "of interest" and "indistinguishable." Here we mean the operations and relations we have defined on the Whole Numbers: successor, addition, and multiplication. We will also include the equals and less-than relations.

Theorem 6.27: The set $\{E(n, 0) \mid n \text{ is a whole number}\}$ with the operations defined above is isomorphic to the Whole Numbers, using the mapping $f(n) = E(n, 0)$. Recall that $E(a, b)$ is the equivalence class containing the pair (a, b)

Proof: The function is onto by definition. It is 1–1 since if $n \neq m$ then $n + 0 \neq 0 + m$ and so $(n, 0)$ is not equivalent to $(m, 0)$ and so $E(n, 0) \neq E(m, 0)$.

(a) Successor is preserved.

$f(n') = E(n', 0) = E((n, 0)') = (E(n, 0))' = (f(n))'$.

(*b*) Addition is preserved.

$f(n + m) = E(n + m, 0) = E((n, 0) + (m, 0)) = E(n, 0) + E(m, 0) = f(n) + f(m)$

(*c*) multiplication is preserved.

$f(n \cdot m) = E(n \cdot m, 0) = E((n, 0) \cdot (m, 0)) = E(n, 0) \cdot E(m, 0) = f(n) \cdot f(m)$

(*d*) less-than is preserved. Suppose n, m are Whole numbers and $n < m$ so there is a Natural Number k such that $n + k = m$. Then $f(m) = f(n + k) = E(n + k, 0) = E(n, 0) + E(k, 0) = f(n) + f(k)$, so $f(n) < f(m)$ since $f(k) = E(k, 0) > E(0, 0)$.

(e) equality is preserved. Let $n = m$ be whole numbers. Then $E(n, 0) = E(m, 0)$ since $(n, 0)\ \alpha\ (m, 0)$. i.e. $n + 0 = m + 0$. Therefore $f(n) = f(m)$.

QED

What we now have is a new set, the Integers, that have, as a subset, something that acts just like the Whole Numbers. We will now explore the properties of the Integers. But first we need a more helpful set of symbols to represent them. Since we are abandoning the Whole numbers we won't need their symbols anymore and we will simply use 0 to represent $E(0, 0)$, 1, to represent $E(1, 0)$, etc. and will also use –1 to represent $E(0, 1)$, –2 for $E(0, 2)$, etc. Note that the symbol –, called *negative*, as used here is unrelated to any notion of subtraction. We need one more theorem to justify our change in notation.

Theorem 6.28: Let E(m, n) be in $\mathbb{Z}$. Then exactly one of the following is true: (m, n) α $(0, 0)$, (m, n) α $(a, 0)$ for some Natural Number a, or (m, n) α $(0, b)$ for some Natural Number b.

Proof. This is a consequence of the Law of Trichotomy for Natural Numbers and the rule of cancellation. Moreover it *is* the law of trichotomy for Integers.

Corollary 6.29: The rule of Trichotomy for < holds in $\mathbb{Z}$.

Below is a visualization of part of the mapping from pairs to integers. Note that each diagonal of the figure has equivalent values. In other words, an equivalence class is represented by a diagonal here. In effect we take the first row and the first column as the standard representative elements of the integers. But you can start at any entry and work right (or upwards) for increasing values and down (or left) for decreasing ones.

		Value of m										
	(m, n)	0	1	2	3	4	5	6	7	8	9	10
V	0	0	1	2	3	4	5	6	7	8	9	10
a	1	-1	0	1	2	3	4	5	6	7	8	9
l	2	-2	-1	0	1	2	3	4	5	6	7	8
u	3	-3	-2	-1	0	1	2	3	4	5	6	7
e	4	-4	-3	-2	-1	0	1	2	3	4	5	6
	5	-5	-4	-3	-2	-1	0	1	2	3	4	5
of	6	-6	-5	-4	-3	-2	-1	0	1	2	3	4
	7	-7	-6	-5	-4	-3	-2	-1	0	1	2	3
n	8	-8	-7	-6	-5	-4	-3	-2	-1	0	1	2
	9	-9	-8	-7	-6	-5	-4	-3	-2	-1	0	1
	10	-10	-9	-8	-7	-6	-5	-4	-3	-2	-1	0

In the set of Integers, elements of the form E(a, 0) are called *non–negative.* If $a \neq 0$ such an element is called *positive*. Likewise elements E(0, a) are called *non–positive,* or *negative* if $a \neq 0$. A positive number is said to have positive (or +) *sign* and a negative number is said to have negative (or –) *sign*.

Theorem 6.30: The less than relation $<$ is well defined in the Integers. That is to say, the definition does not depend on which representation of an Integer is used.

Proof: Let a and b be Integers with $a < b$ and let n be a Natural Number n such that $a + n = b$. Let $a = E(w, x)$, $b = E(y, z)$ and $n = E(n, 0)$. Then $E(w, x) + E(n, 0) = E(y, z)$.

Let (r, s) be an arbitrary element of $E(n, 0)$, so $(r, s)\ \alpha\ (n, 0)$, or $r = s + n$.

Expanding the equation above we see that $E(w + n, x) = E(y, z)$, so $(w + n, x)\ \alpha\ (y, z)$, so $w + n + z = x + y$.

If we add s to both sides of this last equation we get $w + s + n + z = s + x + y$, or $w + r + z = s + x + y$. But then $(w + r, x + s)\ \alpha\ (y, z)$. So $((w, x) + (r, s))\ \alpha\ (y, z)$. Similarly, if we use different representations of a and b we get the same result since addition is well defined. QED

Theorem 6.31: $\leq$ is well–defined.

Proof: Exercise.

Theorem 6.32: The $\leq$ relation is reflexive in $\mathbb{Z}$. i.e. $x \leq x$ for any Integer x.

Proof: The value 0 is the additive identity and equality is reflexive.

Theorem 6.33: The $\leq$ relation is antisymmetric in $\mathbb{Z}$. If $x \leq y$ and $y \leq x$ for Integers x and y, then $x = y$.

Proof: Let m be a Whole number such that $x + m = y$. Let n be a whole number such that $y + n = x$. Then $x + m + n = y + n = x$. By cancellation $m + n = 0$ for whole numbers m and n. But this is only possible if m and n are 0 by the exercise following Theorem 5.15. QED

Theorem 6.34: The $\leq$ relation is transitive in $\mathbb{Z}$.

Proof: Exercise.

Theorem 6.35: The $<$ relation is transitive in $\mathbb{Z}$.

Proof: Exercise.

Theorem 6.36: The relation $\leq$ is a *total ordering* on the Integers: If x and y are Integers then either $x \leq y$ or $y \leq x$.

Proof: Apply the rule of trichotomy for $<$.

Theorem 6.37: If x, y, and z are Integers with $x \leq y$ then $x + z \leq y + z$.

Proof: Exercise.

Theorem 6.38: If x and y are integers and $0 < x$ and $0 < y$ then $0 < x \cdot y$.

Proof: Note that x and y are Natural Numbers here. Therefore the product is a Natural Number. QED

Exercise: If $x + r < y + r$ then what is the relationship between x and y? Does this depend on r? Can you state and prove a theorem?

Note that Theorem 5.17 is not true in the Integers. There are lots of non–empty sets with no least element, for example, the negative Integers. However, you can prove, with a similar argument that any set of Integers that is "bounded below," has a least element.

Definition 6.39: A subset A of $\mathbb{Z}$ is bounded below in $\mathbb{Z}$ if there is an element of $\mathbb{Z}$ that is less than every element of A.

Theorem 6.40: Any subset A of $\mathbb{Z}$ that is bounded below, has a least element.

Proof: Exercise.

You can give a similar definition of bounded above. What theorem can you prove about sets that are bounded above?

Generalization and the Nature of Mathematics – Foreshadowing Abstract Algebra

We have seen earlier that abstraction is key to understanding what mathematics is and how mathematicians think. One thing we like to do is to take some system and generalize it. The integers are an interesting system. By extracting some interesting features from the integers we can create new mathematics. One way to generalize integers is to focus on the properties of the operations more than the elements.

Definition 6.41: A **Group** is a set G, together with an operation $\cdot$, such that (identity) there is an element, *e*, of G such that $e \cdot g = g \cdot e = g$ for every g in G,

(associative) for all *a*, *b*, *c* in G, $(a \cdot b) \cdot c = a \cdot (b \cdot c)$.

(inverses) for any a in G, there is an element a^{-1} in G such that $a \cdot a^{-1} = a^{-1} \cdot a = e$.

Note that we could restate this definition as a set of axioms.

By this definition, the integers are a group, with 0 being the identity element and + being the operation. There is a huge literature about groups and also various generalizations and specializations of the idea of a *Group*. It is one of the first structures that you study in *Abstract Algebra*. If the operation is also commutative, then the group is called Abelian, after the Norwegian mathematician Niels Henrik Abel (1802 – 1829).

Another generalization of the Integers is that of a **Ring**. This requires a second operation with certain properties. I suggest you explore it a bit.

There are a number of things that excite mathematicians. One is new abstractions, since they lead to new kinds of mathematics. One way to think about this is that mathematics is a vast space in which interesting terrain is described by various abstractions. One of the more interesting things about the terrain is that the abstractions overlap in different ways. One of the most exciting aspects of this is that one of these subfields, say Group Theory, will develop its own way of proving theorems from the axioms. The developers were led to these methods because they were *constrained* to think in a certain way by the axioms and definitions. Some people believe that constraint inhibits creativity, but there is a lot of evidence of the contrary. You yearn to break out of the constraints so you come up with clever ways to think. This is *creativity under constraint*, a favorite topic of mine.

Other mathematicians in other fields of mathematics will, then, look at the proof methods and will be inspired to apply them in their own work. Nothing is quite as exciting to a mathematician than finding a new way to prove things.

Since I would like to stop typing out *mathematics* at every use, it is worth noting, that it is a plural word. The singular is mathematic. There are a lot of mathematics. Group Theory is one. The abbreviation of mathematics is maths, properly, though in the United States and some other places it is written math. In Britain, a student studies maths, not math. I will probably use both abbreviations interchangeably. It is also worth noting that maths are so broad that there is no generally accepted definition of the word mathematics. Mathematicians study anything involving definable structure and relationship.

There are two general classes of mathematics and mathematicians: *pure* and *applied*. Pure mathematicians seek to extend mathematics itself, discovering new mathematical truths. Applied mathematicians, on the other hand, would rather take existing knowledge and do something useful in the world with it. It normally requires specialized knowledge about topics other than maths: science, or engineering, perhaps. The boundary isn't hard and fast, however. Pure mathematicians are happy when their results have a use, and sometimes discover uses themselves. Applied mathematicians do, in fact, discover new truths. But the normal working processes are a bit different. The applied mathematician starts with a problem in the world that needs solving and tries to apply maths. The pure mathematician seeks the unknowns in mathematics itself.

Note on Notation. Up to now we have seen a notation like m' only to represent the successor of a whole number m. But it has a more general use as well, which we shall adopt. Sometimes we want to talk about two values. We might represent them as m and m', rather than x and y, for example, without implying any particular relationship between them. This can be a pitfall for the novice, of course. Mathematicians can use the same notations to represent different things, so you need to be aware of the context.

Applied Mathematics: A Simple Example

All of us are applied mathematicians at some level. We exchange and keep track of money. We estimate distance and time in our travels, etc. Of course, applied mathematics goes beyond this but we can all use mathematics to help us answer some questions that arise in our lives.

Curve Matching is a slightly more sophisticated application that can be useful to anyone interested in science.

Suppose you want to know the temperature of an open vent in a volcano. Suppose your only tools are an ordinary thermometer and mathematics. You can approach the vent, but not too closely as the danger of ground collapse increases as you get closer.

One way to do this, at least approximately, is to use *curve matching* and *extrapolation*. Curve matching tries to find a mathematical function that matches actual measurements. Once you have a curve that seems reasonable, you use the values of the curve to get an estimated value beyond the range that you actually measured. This last is extrapolation.

Here we will give a few simple examples: linear, quadratic, and exponential matching and extrapolation.

Linear: Since two distinct points determine a straight line, let us take two measurements near the volcano vent and do some computations. Let's first determine a simple axis along which to measure. At some point reasonably near the vent (reasonable in terms of safety) where the heat from the vent can already be felt, mark a zero point. Define the vent location itself as unit four. This is completely arbitrary, but we now have a simple scale. Suppose the temperature at point 0 is 50; the temperature scale doesn't matter. Suppose that at point 2 the temperature is 200. We now have two points: (0, 50) and (2, 200). The straight line through these two points has slope 75 = (200-50)/(2-0). The linear function that is defined by these two points is $t(x) = 50 + 75x$. We now have a curve that seems to match the temperature gradient. The match is very approximate, of course, since we made an assumption about linearity and the assumption may not be valid. But if we extrapolate on this curve to point 4 we get a value of 350. So, our approximation of the temperature at the vent is 350 units.

Quadratic: We can probably get a better approximation if we notice that the temperature seems to rise faster than linearly. If you halve the distance the temperature seems to more than double. Let's try the same trick using a quadratic approximation. Three distinct points not on a line determine a parabola. The general formula for such a quadratic curve is $f(x) = a + b \cdot x + c \cdot x^2$, for some constants a, b, and c that we need to determine. Let's go back to our measurements. Let's measure the temperature at point 1 on our scale (one fourth of the way from our arbitrary point 0 to the vent). Suppose the temperature is 100. We now have three points to work with: (0, 50), (1, 100), and (2, 200). Using just the first point we see that $f(0) = a + b \cdot 0 + c \cdot 0^2$, so a must be 50. We need to determine both b and c.

$100 = f(1) = 50 + b \cdot 1 + c \cdot 1^2$, so $b + c = 50$, or $b = 50 - c$.

$200 = f(2) = 50 + b \cdot 2 + c \cdot 2^2$, so $2 \cdot b + 4 \cdot c = 150$. Simplifying $b + 2 \cdot c = 75$.

If we substitute the value for b $(50 - c)$ into this equation we get $50 - c + 2 \cdot c = 75$, so $c = 25$.

But then the first equation says that b is also 25.

Thus our quadratic function is $f(x) = 50 + 25 \cdot x + 25 \cdot x^2$. And note that this parabolic curve contains all three of our points. Now if we extrapolate to point 4 we get $f(4) = 50 + 25 \cdot 4 + 25 \cdot 16 = 550$. This is quite a bit higher than our linear approximation.

Exponential: Finally, lets try to match an exponential function to our data: $g(x) = a \cdot 2^{b \cdot x}$. We will use the same three points to determine a and b. The first point, (0, 50), says that $50 = g(0) = a \cdot 2^0 = a$, so $a = 50$.

We only need one additional point to determine b, so which shall we use. Probably our result will be more accurate if we choose the measurement closer to the vent since we won't need to extrapolate so far.

Thus $200 = g(2) = 50 \cdot 2^{2 \cdot b}$, or $4 = 2^{2 \cdot b}$. If we take the logarithm (base 2) of both sides of this we get that $2 = 2 \cdot b$ or that $b = 1$. Thus, our function is $g(x) = 50 \cdot 2^x$. Note that this curve contains both (0, 50) and (2, 200). We haven't used the middle estimation point at all here.

If we extrapolate with this function to point 4 we get $g(4) = 50 \cdot 2^4 = 800$ units.

I made up all these numbers for ease of computation, but it turns out that the temperature at a vent, measured by more sophisticated means is between 700 and 1200 degrees Celsius.

Actually, in this case the middle measurement, (1, 100), gives the same function, but that is just serendipity. It just happens that this exponential curve and the parabola we computed earlier (quadratic function) happen to contain these three points. I chose the points for ease of computation only, just as I decided to use a base of 2 for the exponential function, rather than the more common 10, or e, Euler's constant.

Also, with several measurements, one can use more sophisticated methods of curve matching in which you find a "best fit" to a set of data rather than trying for an exact fit. See *Least Squares Approximation* (+).

And if you combine science, especially thermodynamics, with mathematics you can do even better.

Subtraction

Our motivation for developing the Integers was that we could not solve certain simple equations in the Whole Numbers. Equations like $x + 5 = 3$ have no solution with x as a Whole Number. The ancient Greeks thought such equations were absurd.

Definition 6.42: If x and y are Integers, with E(a, b) = x and E(c, d) = y, we define the *difference* of x and y as $x - y$ = E(a +d, $b + c$). We say that y is *subtracted* from x. We call x the *minuend* and y the *subtrahend*. Note that the ordered pair (c, d) was simply reversed and the sum applied.

So 3 – 5 is E(3, 0) – E(5, 0) = E((3, 0) + (0, 5)) = E(3, 5) = E(0, 2) = –2 as we expect.

Theorem 6.43: Subtraction is well–defined.

Proof: Let (A, B) α (a, b) and (C, D) α (c, d). Then E(A, B) – E(C, D) = E($A + D$, $B + C$).

We need to show that this is equal to E($a + d$, $b + c$), or that:

$A + D + b + c = a + d + B + C$. But $A + b = a + B$ and so

$A + D + b + c = a + D + B + c$ by substitution. But $C + d = c + D$, so this last is

$a + d + B + C$. This completes the proof of equivalence. QED

Theorem 6.44: Subtraction is not commutative. Note that this doesn't mean that $a - b$ is never the same as $b - a$. It means that there is at least one example where it fails.

Proof: In particular $5 - 2 \neq 2 - 5$. $(5, 0) - (2, 0) = (5, 0) + (0, 2) = (5, 2) \, \alpha \, (3, 0)$. But $(2, 0) - (5, 0) = (2, 0) + (0, 5) = (2, 5) \, \alpha \, (0, 3)$. But $(3, 0)$ is not equivalent to $(0, 3)$ since $3 + 3 \neq 0 + 0$. QED

Theorem 6.45: Subtraction is not Associative.

Proof: Exercise.

Theorem 6.46: Multiplication is distributive over subtraction: $x \cdot (y - z) = x \cdot y - x \cdot z$ for Integers x, y, z.

Proof: Partial. Let $x = \mathrm{E}(a, b)$, $y = \mathrm{E}(c, d)$, and $z = \mathrm{E}(e, f)$. Then $x \cdot (y - z) = \mathrm{E}(a, b) \cdot (\mathrm{E}(c, d) - \mathrm{E}(e, f)) = \mathrm{E}(a, b) \cdot (\mathrm{E}(c, d) + \mathrm{E}(f, e)) = \mathrm{E}(a, b) \cdot \mathrm{E}(c + f, d + e) = \mathrm{E}(a \cdot (c + f) + b \cdot (d + e), a \cdot (d + e) + b \cdot (c + f))$. Finish this proof by doing a similar computation on the right hand side (RHS) and showing the results are equivalent. You may take advantage of the rules for the Whole Numbers, of course, since a, b, etc. are Whole Numbers in these computations. This completes the proof of left distributivity. The proof for the right distributive rule is similar. You may also apply the commutative law to prove the right distributive law.

A Note on Proof

You may have noticed that our methods of proof have changed from when we considered the Natural Numbers and Whole Numbers. Those proofs were nearly all based on induction. We haven't used that at all here. However, we have based our definitions on the Natural Numbers and the proofs now are built up from properties that we proved earlier about them. This is the power of abstraction.

On the other hand, the proofs here aren't very exciting. And if they are a bit tedious, they are tedious in a special way. We don't have to separate out cases of the non–negative numbers from the negative ones for any of our operations or proofs. Since we gave a unified definition of the Integers, even if seemingly strange, a great many difficulties were saved. The problem with case analysis in a proof is that you might miss one and then your proof fails. Had we defined the integers as the union of the whole numbers with something that looks like a copy of the Natural Numbers, we would have had a complex situation in the definitions of our operations and in the proofs of their properties. That wouldn't be very exciting either.

Substitution and Cancellation Laws in $\mathbb{Z}$

Theorem 6.47. If a and b are Integers then $a + x = b$ has a unique solution in the Integers.

Proof: The solution is $(b - a)$ as may be easily shown. QED

Exercise: Show that the solution in Theorem 6.47 is unique.

Corollary 6.48: If a and b are Integers then $a - x = b$ has a unique solution in the Integers.

Proof: Similar to the above.

Corollary 6.49: If a is an Integer then $a + x = 0$ has a unique solution. The solution is called the *additive inverse* of a. It is also called the *negative of a*. We noted earlier that the Integers form a Group. This result establishes the fact. Note that $E(r, s) = -E(s, r)$ as has been noted previously.

Note, however, that we cannot say the same thing about multiplication. For example, $2 \cdot x = 1$ has no solution in the Integers. We would need to extend again to have a system in which we can always solve such equations.

Theorem 6.50: If a is a Natural Number (i.e. is positive) and b, and c are Integers with $b < c$, then $a \cdot b < a \cdot c$.

Proof: If n is a Natural Number with $b + n = c$ (the definition of <, 6.24), then $a \cdot b + a \cdot n = a \cdot c$ and $a \cdot n$ is a Natural Number. QED

Exercise: Find and prove the equivalent theorem for the cases (i) $a = 0$, (ii) $a < 0$. Because of these theorems, multiplication is said to be *monotonic* over $\mathbb{Z}$.

Definition 6.51: If $x \in \mathbb{Z}$, define $|x| = x$, if x is a Whole Number, and $-x$ otherwise. $|x|$ is called the *absolute value* of x.

Theorem 6.52: The equation $a \cdot x = 1$ has no solution in $\mathbb{Z}$ unless $|a| = 1$.

Proof: Suppose $a \neq 1$. If $a = 0$ then $a \cdot x = 0$ for any x, so there is no solution. If $a > 1$, then $a \cdot x > 1 \cdot x = x$ for any positive x. If x is negative, on the other hand, then $a \cdot x < 0$, so is not 1. This proves the case for $a > 1$. If $a < -1$, then for any positive x, $a \cdot x < 0$, so is not 1 and for any negative x, $a \cdot x > 1$. QED

Lemma 6.53: If $x = 0$ then $-x = 0 - x = 0$.

Proof: Exercise.

Lemma 6.54: Negative of a product. $-(x \cdot y) = (-x) \cdot y$ if x, y are in $\mathbb{Z}$.

Proof: Let $x = \mathrm{E}(a, b)$ and $y = \mathrm{E}(c, d)$. Then $x \cdot y = \mathrm{E}(a \cdot c + b \cdot d, a \cdot d + b \cdot c)$.

Then, $-(x \cdot y) = \mathrm{E}(a \cdot d + b \cdot c, a \cdot c + b \cdot d)$. By definition, $-x = \mathrm{E}(b, a)$. And $(-x) \cdot y = \mathrm{E}(b \cdot c + a \cdot d, b \cdot d + a \cdot c)$. Now apply commutative law of addition in the Whole Numbers. QED.

Theorem 6.55: If a and b are in $\mathbb{Z}$, and if $a \cdot b = 0$ then a or b (or both) is zero.

Proof: By the two previous lemmas we may assume without loss of generality (WLOG) that both a and b are non negative since if they are not then their absolute values will still have product 0. Let $a = \mathrm{E}(x, 0)$ and $b = \mathrm{E}(y, 0)$ for Whole Numbers x and y. Then $a \cdot b = \mathrm{E}(x \cdot y, 0)$. If this is zero then $x \cdot y = 0$ and so the result follows from Theorem 5.33 since if both x and y are equal to 1 then so is their product, but if either is greater than 1 the product is greater than the other. So if both are greater than zero then their product is also. QED

The following exercises depend on the corresponding laws of Whole Numbers.

Exercise: Prove the substitution law for addition in $\mathbb{Z}$. If a, b, and x are Integers and a = b then a + x = b + x.

Exercise: State and prove the substitution law of multiplication in $\mathbb{Z}$.

Theorem 6.56: Let a, b, c be in $\mathbb{Z}$. If $a + c = b + c$ then $a = b$. This is the *law of cancellation for addition*.

Proof. Add $-c$ to both sides (substitution law) and get $a + 0 = b + 0$. The result follows. QED

Theorem 6.57: Let a, b, c be in $\mathbb{Z}$. If $c \neq 0$ and $a \cdot c = b \cdot c$, then $a = b$. This is the *law of cancellation for multiplication.*

Proof: Proof by Contradiction. Assume $a \neq b$. Then $a < b$ or $a > b$. If the former, then there are two cases. If $c > 0$ then $a \cdot c > b \cdot c$ so they are not equal. The other cases (3) are similar.

Theorem 6.58: If a, b, c are positive Integers (i.e. Natural Numbers) and $a \cdot b = c$, then $a \leq c$ and $b \leq c$.

Proof. By symmetry it is enough to show that $a \leq c$. But $1 \leq b$, so $a = 1 \cdot a \leq a \cdot b = c$. (see Theorem 5.31) QED

Theorem 6.59: Any non empty set of Integers that is bounded below has a least element. Any non empty set of Integers that is bounded above has a maximum element.

Proof. Similar to Theorem 5.17 and its corollary. OR. Apply a bijection that maps the set into some set of Natural Numbers. For example. If set S is bounded below by b then consider $S' = \{ x \mid x = b + s \text{ for some } s \text{ in } S \}$. Then S' has a least element t. But $t = b + s$ for some s in S. This is the least element. Showing that is straightforward.

Integer Division and Remainders

We would like to define a new operation, *division,* in the integers. It isn't possible, of course, since, for example, $10 \div 3$ has no real meaning in the integers. We can use cancellation, of course, since when $ab = ac$, then $b = c$ if $a \neq 0$, but that isn't quite the same thing. We note, however, that $9 \div 3$, according to what you learned in school, is 3 and so if $10 \div 3$ were to have a meaning then it would be something like "a bit more than 3".

Definition 6.60: Let a and b be integers, with $b \neq 0$. We say that b *divides* a provided that there is an integer m, such that $b \cdot m = a$. In that case we write $b \mid a$ and say that b is a *factor* of a.

Note that this isn't an operation. It is a relation. It is a statement that can be true or false. It is false that $3 \mid 10$, but true that $5 \mid 10$. If d is a factor of a, then, according to

Theorem 6.58, $1 \leq |d| \leq |a|$. Note that 1 and x always divide x. The set of integers that divide a given number is a finite set. The factors of x, other than 1 and x, are called non–trivial factors. So 5 is a non–trivial factor of 15, but 15 is a trivial factor of 15.

If a does not divide b we sometimes write $a \nmid b$. Thus $3 \nmid 10$. In fact we generally indicate that some relation does not hold by "slashing" its symbol. For example, $x \neq y$.

We can, however, jointly define a pair of operations once we prove the following theorem.

Theorem 6.61: If a and b are any integers, with $b \neq 0$, then there exist unique integers q and r such that $a = b \cdot q + r$, with $0 \leq r < |b|$. In such an operation, the value a is called the *dividend*, and b is the *divisor*. Moreover, the value q is called the *quotient*, and we will write it $a\backslash b$, using a backslash. The value r is called the *remainder*, and we denote it with $a\%b$, or *a mod b*. *Mod* is short for modulo.

The names *div* for quotient and *rem* for remainder are also common, especially in computer programming languages. Note that, while these operations might be available in your favorite computer language, they might not give the same values, especially if a or b is negative.

Proof: First we will assume that $a \geq 0$ and $b > 0$. Finding two numbers that satisfy the equation $a = b \cdot q + r$ is easy, of course. Just let $q = 0$ and $r = a$. But that doesn't necessarily satisfy $0 < r < |b|$, nor does it give any notion that the two values are unique. However, if it turns out that $r < b$ we have the existence, but not uniqueness. If that is not the case, starting with this q and r, perform the following repeatedly. Replace q by $q + 1$ and r by $r - b$.

Then $b \cdot (q + 1) + (r - b) = b \cdot q + b + r - b = b \cdot q + r = a$. So, we have the required equation, but for a larger q and a smaller r. Repeat this until we have an $r < b$. In the previous step, r was not less than b so then it must satisfy $0 \leq r < b$. We have established the existence.

To show that the final q and r are unique, suppose that for some q_1 and r_1, we also have $a = b \cdot q_1 + r_1$ with $0 \leq r_1 < b$, which is equivalent to $-b < -r_1, \leq 0$. Adding that to the inequality $0 \leq r < b$, we get $-b < r - r_1 < b$, or $0 \leq |r - r_1| < b$. From the two equations $a = b \cdot q + r$ and $a = b \cdot q_1 + r_1$, we get that

$b \cdot (q_1 - q) = r - r_1$, and so b divides $(r - r_1)$. But then, if $r - r_1$ isn't zero it satisfies $r - r_1 \geq b$, contradicting the inequality we just established.

To finish we must cover the cases in which a or b is negative. Suppose that we have our solution q, and r for the situation $c = |a|$ and $d = |b|$, $c \geq 0$ and $d > 0$.

First, suppose $b < 0$, $d = -b$ with $a > 0$. Then $c = d \cdot q + r = (-d) \cdot (-q) + r = b \cdot (-q) + r$. So the quotient is -q and the remainder is r.

Next suppose that $a < 0$, $c = -a$ with $b > 0$. Then $a = -c = d \cdot (-q) - r = d \cdot (-q - 1) + d - r$. The, the quotient is -q - 1 and the remainder is d – r which is positive and < d.

Finally, if both a and b are negative, notice that

$-a = (-b) \cdot q - r$, and so $-a = (-b) \cdot (q + 1) + (b - r)$. QED

We were guided in these transformations simply by the need to keep the remainder non–negative, but less than the divisor.

QED

Examples: $10 = 3 \cdot 3 + 1$ and so $-10 = 3 \cdot (-4) + (-1 + 3)$. The quotient is –4 and the remainder is 2: $(-10) \setminus 3 = -4$ and $(-10)\%3 = 2$. Also $-10 = (-3)(4) + 2$. The quotient is 4 and the remainder 2. And: $10 = (-3) \cdot (-3) + 1$, and $10\setminus(-3) = -3$, with $10\%(-3) = 1$.

As mentioned above, we would like to have a division operation in the Integers, but the fact that $a \cdot x = b$ has an Integer solution only in certain circumstances makes this impossible. We do have a cancellation law, of course, which is related to, but not the same as, division. We need to extend our number system again before we can define true division.

But first, the very impossibility of solving such equations in general opens new possibilities to an inquiring mind. This is especially useful in Number Theory. We use the remainders to define new kinds of operations: modulo operations.

Suppose we take a set of integers such as $I_6 = \{0, 1, 2, 3, 4, 5\}$, which has six elements. Let's define an operation $+^6$ as follows. If a and b are in I_6 then $a +^6 b = (a + b) \% 6$. Then $2 +^6 5 = (2 + 5) \% 6 = 7 \% 6 = 1$. This new operator is called the "sum modulo 6", or "sum mod 6". Note that since all of the elements of I_6, considered as integers are non–negative, the results of this operation are always in I_6, so that I_6 is *closed* (see pg. 49) under sum mod 6. We can define subtraction and multiplication similarly. Note that 0 is still an additive identity and 1 a multiplicative identity. This is essentially, clock arithmetic, of course. Note that integer arithmetic in computers in essentially like this but using a set much larger than six.

You can do interesting things in such a system. In particular you can try to solve equations. What is the solution, here, to $5 +^6 x = 2$? Explore this a bit. It can be quite fun.

Note that $\leq$ has no meaning in I_6 nor in similar sets. The number 5 is both bigger and less than three mod 6, for example, at least using our original definition of $<$. Note that $3 +^6 2 = 5$ and $5 +^6 4 = 3$. So each is less than the other.

Definition 6.62: If x is a factor of a, we say that a is a *multiple* of x.

Integers that are multiples of the number 2 are called *even*, the others are called *odd*. The names have nothing to do with temperament, however[1]. Two itself is the only even prime. Zero is considered even. It *is* a multiple of 2, of course.

Lemma 6.63: If x is a whole number then exactly one of x and x' is even.

Proof: $x = 2 \cdot k + r$ with $r = 0$ or $r = 1$. If $r = 0$ then k is even. If $r = 1$ then $x' = 2 \cdot k + 1 + 1 = 2 \cdot (k + 1)$ and so x' is even.

The following isn't quite ready for prime time as we haven't defined / yet. But here we note that \ works just as well as a note below will verify.

Theorem 6.64: If m is any Whole Number, then $\sum\{ n \mid n \in \mathfrak{W}, n \leq m \} = m(m + 1)/2$.

Proof: by induction on m

(a)If $m = 0$ the set consists of 0 alone and the right hand side is also 0 since we multiply by 0.

(*b*) Suppose the statement is true for some Whole Number m. The induction hypothesis is $\sum\{ n \mid n \leq m \} = m(m + 1)/2$. Now consider m'. The left side becomes $\sum\{ n \mid n \leq m' \}$. But since we are now adding in m', this is $\sum\{ n \mid n \leq m \} + m'$. By induction this is $m(m + 1)/2 + m'$ or $m(m + 1)/2 + (m + 1)$.

[1]That is what passes as a joke among mathematicians.

Finding a common denominator (i.e. 2) we see this is $(m(m + 1) + 2(m + 1))/2$. Factoring out $m + 1$ from the numerator we get $(m+1)(m + 2)/2$, which is the right side of the formula with $m+1$ replacing m: $m'(m' + 1)/2$. QED

Note that we have used a restricted form of division. But we haven't left the Integers: $m\cdot(m+1)/2$ is an integer since either m or $m+1$ is even. Therefore the product is even and so $m\cdot(m+1)/2 = m\cdot(m+1)\backslash 2$.

There is a story about the above Theorem, or its formula, at least. When a young boy, the mathematician Leonhard Euler (1707–1783) was thought to be a difficult child in school. It was because he was very smart and his teachers didn't really know how to deal with him and his questions. One day, to try to silence him for a while, the teacher told him to add up the numbers from 1 to 100, thinking it would take a while. The kid immediately announced the answer is 5050. He didn't actually know the formula but reasoned like this: 1 + 99 = 100, but so does 2 + 98, 3 + 97, etc. There are 49 such pairs, with 50 and 100 not paired. So the sum is 49·100 + 150. Or, the alternate way to do it is to take two copies of the numbers, with one reversed and just add corresponding pairs. Each of the 100 pairs adds to 101, but you have to halve the total since you used two copies.

It is my firm belief, supported by anecdotal evidence, that many kids are thought difficult when it is only true that they are smart, but bored, in school.

There is a complete and very interesting field of mathematics called Number Theory(+) that has many marvelous theorems, but, even more important, many things that are not yet known. In the eighteenth century important advances were made in Number Theory. In that field you study other things than modulo arithmetic, such as prime numbers. That takes you to encryption, which is a vital topic in the early 21st century. The study in interesting both technically and from a public policy standpoint. While I was writing this book a new largest known prime was found: $p = 2^{74,207,281} - 1$. It has about 22 million digits. It is unknown how many smaller primes exist. If p did have a factor, the largest could have as many as about 11 million digits. Primes with this form, one less than a power of 2 are called Mersenne primes(+), named after a 17th century monk, Marin Mersenne, who studied them. Not all numbers of the form $2^n - 1$ are prime, however, even when n is prime. Note that 74,207,281 is, itself, a prime. Here are a few Mersenne primes with an interesting property: 3, 7, 127. Note that 7 is $2^3 - 1$ and 127 is $2^7 - 1$. It is also true that $2^{127} - 1$ is prime, but has too many digits to list here. What is it about these numbers that relates them in this way? How often does it hold? Is anything known about it? Can you discover anything? Note that 31 is a Mersenne prime as is $2^{31} - 1$. I think the problems posed in the questions above are very difficult.

As a simple example of a theorem from Number Theory, we can prove the following.

Theorem 6.65: There are an infinite number of primes.

Proof: By contradiction. Suppose there are a finite number of primes. Suppose they form the set P = $\{p_1, \ldots, p_n\}$. Now the number q that is one more than the product of all of these is (a) larger than any of the p's, and (b) itself prime. The latter is true because the remainder of q mod $p_i = 1$ for any i. So the value q is a prime not in P, which is a contradiction. QED

See whether you can determine whether the following is a theorem or not.

Proposition: If $\{p_1, \ldots, p_k\}$ consists of the smallest k primes (i.e. $\{2, 3, 5, \ldots, p_k\}$), then their product plus 1 is also prime. For example $2 \cdot 3 \cdot 5 + 1 = 31$ which is prime.

Since there are many open questions in Number Theory it is a fruitful field of study for a new mathematician.

It isn't our purpose in this book to give a complete picture of Number Theory, but we can explore a few important results. The theorem, above, defining the quotient and remainder is one.

A number that divides both a and b is called a *common divisor*(+) of a and b. The greatest of these is called the greatest common divisor, or GCD(a, b). For example GCD(20, 30) = 10. Euclid's Algorithm(+) (not explored here, but you can search for it) can be used to compute the GCD of any pair of Integers. Note that the GCD(a, b) > 0, since if a negative value is a common divisor then so is *its* negative which is greater but also positive.

Theorem 6.66: (Bézout's Identity) Let a and b be any non-zero integers. Let GCD(a, b) = d. Then there exist integers x and y such that $ax + by = d$.

Example. The GCD(4, 6) is 2, as is easy to see by inspection. Also $(-1)\cdot 4 + (1)\cdot 6 = 2$. However, try to convince yourself that there are no integers x, and y, such that $4x + 6y = 1$.

Proof: Let a and b be Natural Numbers. Let $B(a, b) = \{ ax + by \mid x, y \text{ are integers with } ax + by > 0\}$. This set is non–empty since $a\cdot a + b\cdot b > 0$. Let d be the smallest element of $B(a, b)$. Then $d = ax + by$ for some x and y. We must now show that $d = \text{GCD}(a, b)$.

Since $a = d \cdot (a\backslash d) + a\%d$ we have:
$a\%d = a \cdot 1 - (ax + by)\cdot(a\backslash d) = a \cdot 1 - ax\cdot(a\backslash d) - by\cdot(a\backslash d) =$
$a \cdot (1 - x\cdot(a\backslash d)) - b \cdot (y \cdot (a\backslash d))$. Thus, if $a\%d > 0$ it is in $B(a, b)$ and so $d \leq a\%d$. But, by definition, $a\%d < d$. The only alternative then is $a\%d = 0$, so d divides a. Similarly, $d \mid b$ and so d is a common divisor of a and b. We are getting closer to showing $d = \text{GCD}(a, b)$. Now let c be any common divisor of a and b. Then c divides $ax + by = d$, so $c \mid d$. But then $c \leq d$. QED

Corollary 6.67: If a and b are non–zero integers, then GCD(a, b) is the smallest number of the form $ax + by$, where x and y are arbitrary integers. This is a consequence of the proof of the theorem.

A Note on Mathematical Insight

I think that Bézout's Identity is a wonderful thing. It feels to me like it took marvelous insight for Bézout to discover it. My own insight is in different fields and I admit that I don't have much insight into Number Theory. My main field was Analysis (think Calculus and its extensions) with Point Set Topology a close second. I also liked Logic and Set Theory when I was a practicing Mathematician many years ago. But Number Theory and Abstract Algebra escaped me. So, insight into mathematics isn't necessarily a general thing, but rather specific.

But how did Étienne Bézout (1730–1783) go about discovering this? Actually there is evidence that he wasn't first, but still, someone did. Given the statement of it, it seems fantastical. But I suspect that the originator, Bézout or another, experimented a lot with similar ideas before he or she got there. But experimentation in mathematics is different from that in Physics or Biology. It doesn't involve the real world. It wasn't arrived at by observing sheep pens, for example, but by looking a various equations and manipulating them, looking for patterns.

But the *insight* into what *might be true* was the essential thing. It is one thing to prove a theorem that you already know to be true, since it was proved in the past. It is another thing to be given a statement and asked to explore its truth. Someone else thinks it is worth your time to examine. But the step beyond that is to find the things the aren't known at all. To formulate a *Maybe Theorem* and then to prove it true or show why it isn't. A kind of groping for the truth.

An important and unsolved problem in Number Theory is Goldbach's strong conjecture (more often, simply "Goldbach's conjecture") which says that every even number larger than 2 can be written as the sum of two prime numbers.

Another theorem was only recently proven. Fermat's Last Theorem has an interesting story. It states, quite simply that $x^n + y^n = z^n$ has no Natural Number solutions if $n > 2$. That is, no Natural Number choice for x, y, and z will make this true. If $n = 2$ there are infinitely many solutions, such as (3, 4, 5). If $n = 1$ solutions are trivial to find, of course. The case for $n > 2$ was an open question from 1637 until 1994 and it's study was important in the development of Number Theory. The story is that Pierre de Fermat wrote the theorem in the margin of a book he was reading, as was his habit, but noted that there wasn't room to write his "elegant" proof.

The following if one of the most important theorems in math history.

Theorem 6.68: (Euclid's Lemma) If p is a prime number and $p \mid a{\cdot}b$, then $p \mid a$ or $p \mid b$ (or both).

Examples. 3 is prime and it divides 24, which is $4 \cdot 6$. But 3 divides 6. On the other hand, if p isn't prime we can't conclude such a thing. 12 also divides 24, but it divides neither 4, nor 6.

This theorem is fundamental in Number Theory and establishes that prime numbers are special.

Definition 6.69: If GCD(a, b) = 1 then a and b are called *relatively prime*. Note that if p is a prime number and p doesn't divide a, then a and p are relatively prime. The converse is also true. If p is prime and relatively prime to a, then p doesn't divide a.

Example: Fourteen and fifteen are relatively prime. Fourteen and sixteen are not.

We prove a more general form of Euclid's Lemma.

Theorem 6.70: If $p \mid a \cdot b$ but p and a are relatively prime then $p \mid b$.

Proof:

> If p and a are relatively prime then by Bézout's Identity, for some x and y, $px + ay = 1$. Multiply both sides by b to get $bpx + bay = b$. Since $p \mid ab$ it divides both terms on the left of this equality, hence it divides the sum and therefore divides b. QED.

Euclid's lemma follows as a special case.

The following theorem is another of the most important results of Number Theory. We state it, but leave it for you to explore. Euclid's Lemma is the key.

Theorem 6.71: (**Fundamental Theorem of Arithmetic**). If p is a Natural Number then p is the product of primes in exactly one way if the order of the factors is ignored. Note that factors may be repeated. This is also called the **Unique Factorization Theorem**[(+)].

Examples: $9350 = 2 \cdot 5 \cdot 5 \cdot 11 \cdot 17$. Some of the factors of 9350 are easy to find, since it is even, etc. $21{,}793 = 19 \cdot 31 \cdot 37$ is a bit harder to figure out, though knowing the answer it is easy to check. This is the essence of most encryption algorithms. Large numbers are hard to factor, but if you know the factors it is easy to check.

Often is is helpful to know if a number is evenly divisible by another. There are some interesting tricks. A number expressed in base ten is divisible by 3 if and only if the sum of its digits is also divisible by 3. The same is true for 9. You recognize even numbers and numbers divisible by 5 instantly, of course. See if you can find a rule for divisibility by four. By eight. But these are a simple consequence of the base we normally use. Both two and five evenly divide ten. Both 3 and 9 have remainder 1 when dividing into ten. If we used base twelve, say with "digits" 0, 1, 2, 3, 4, 5, 6, 7, 8, 9, α, and β, then we would recognize that, for example, a number is divisible by 3 if and only if it ends in 3, 6, 9, or 0, but not otherwise. We could also recognize multiples of 2, 3, 4, 6, and 12 instantly just by looking at the last "digit." But we would need to do something else to recognize multiples of 5.

For example, in base twelve, the number twenty-five looks like 21. Oh, and ten looks like α.

Summary So Far

Starting with the Natural Numbers defined via the Peano Axioms we have extended twice and found that the resulting systems have similar, but not identical properties. We have operations like + and · and relations such as ≤ that extend from one system to the other, but not quite identically. For example, there are no additive inverses, necessarily, in $\mathfrak{N}$ or $\mathfrak{W}$ though there are in $\mathfrak{Z}$. Along the way we have, I hope, gained some insight into how mathematicians think and what they do. Much of what they do is tedious and boring, such as the repetitive proofs of many of the theorems about the integers. But in order to make those boring, someone needed the insight to represent the integers as equivalence classes so that the various proofs didn't all require special cases and a lot of creativity. That insight lead to an elegant solution. The one big insight made the rest somewhat routine. The various theorems we have presented give properties of the number systems and their operators and relations and enable useful work. The commutative and associative laws, for example, greatly simplify many computations. Image doing algebra without them.

Note that the three systems we have already developed are quite similar to one another. The elements are discrete. There is no number in any of these systems between 4 and 5. There are only a finite number of values between 2 and 5, or between –5 and 5.

Some of the individual values have interesting properties, such as 0 and 1. Some *kinds* of values have interesting properties, such as prime numbers. Not all of these interesting properties have been discovered. So mathematics isn't a fixed system, all wrapped up. It is a vast space inviting exploration. But explorers need to prepare themselves, gathering the tools of exploration.

As I was writing this book, a fifty year-old problem, the Kadison–Singer problem[*], was solved. It involved both Topology and Analysis, and had implications in applied math as well.

Also note that in defining the Integers, we were able to take account of known laws about the Whole Numbers. Most of our proofs, in fact, were based on properties of Whole Numbers. That was a big advantage of our representation of the Integers as equivalence classes of pairs of Whole Numbers. We shall see this idea again (and again). In developing the Natural Numbers and the Whole Numbers we were, among other things, establishing a solid base on which to stand in developing the Integers.

We will continue to extend our number systems. The remaining three we shall study are very unlike the three we have already covered. And they are very unlike each other in many ways. The base we have established will allow an explosion of possibilities. The tools we have learned will permit us to develop additional tools. And on and on.

Nor will it end at the end of this book. The ideas here only scratch the surface of a small part of that vast space.

What if there *were* a number between 4 and 5? Hmmm...

Before we look at the next extension, the Rational Numbers, however, let's take a short side trip. Earlier we looked at the idea of a Group, which is a set with a closed, associative operation with inverses. The Integers are an example of a group. So are the Integers mod 6 (or other bases). You can even suppose that the Integers inspired the development of Group Theory as a generalization. Likewise a *Ring* is a set with two operations. The first operation is normally called addition and with this operation a Ring is a commutative (Abelian) Group. The other operation, normally called multiplication, must also be associative, but not necessarily commutative. And the multiplication operation needs to distribute over the addition operation. Sound familiar. A Ring, then, is an abstraction of the Integers, with some, but not necessarily all of the properties of the Integers. Having fewer constraints, more things *can* be true in a Ring than in the Integers, but fewer things *must* be true. There turns out to be a huge difference between commutative Rings (in which the multiplication operation is commutative) and the non–commutative Rings. Go study.

So, it seems, that some of the ideas of number have inspired mathematicians to invent more abstract things and to study them. There are even simpler algebraic structures, in fact. A *Semigroup* is just a set with a closed operation, such as the

Natural Numbers with addition. I used to like to study such extremely simple structures; to see what could be proven, that then must be true also of more complex structures based on them.

Later we will look briefly at some other abstractions based on the number systems we are studying here, leading to other kinds of maths entirely.

The Maya numerals are remarkably like our tick marks. Note that they are useful computationally. A "five" and a "six" make up an 11. Here are the first fifteen symbols:

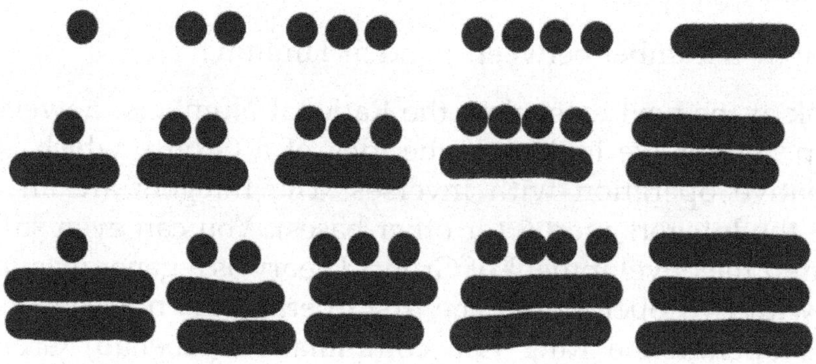

They also had a symbol for zero that was used positionally, like ours. It was drawn as the glyph of a shell. Something like this:

7 The Rational Numbers

So far, the numbers we have considered have been good for counting and even recording debits and credits. But they haven't been so good at measuring. While some things (sheep) are discrete and only come in whole units, some things arise naturally in parts; quantity rather than count. If I have gold bars and want to buy bread it is very difficult. I would need to buy more bread than I could possibly use in trade for a single gold bar. So we need to be able to talk about fractional parts. Historically, *pieces of eight* were silver coins that could be easily divided, conceptually, into eight parts. An old Spanish Dollar equals eight Reals. Later this evolved, possibly in Italy or New York City, into Pizzas of Eight. Very useful for sharing with friends.

Rational Numbers can be expressed as p/q where p and q are integers but $q \neq 0$. They have other representations. One of them is a repeating decimal expansion. Note that the notation p/q does not necessarily imply division. It is just a notation. We could use a comma or a question mark just as well. In fact, it is really just an alternate notation for an ordered pair. Eventually we will see division and then see that the notation we use for the Rationals is suggestive of division.

You probably already know that a given Rational Number can be represented different ways. So 10/3 is the same value as 30/9. Again this suggests, well, something; division perhaps, or cancellation. The name *rational* comes from the word ratio. It isn't about mental states, though we will also, eventually, see irrational numbers. That name is a bit of a joke, I think. See, mathematicians can be very funny. We shall see. In fact, though, the word "rational" comes from a Latin root that means calculation. The Pythagoreans(+) of Greece were the original "rationalists," based on their belief that all measurement could be reduced to comparison of integer values.

But first we need to define the rationals and to show that $\mathfrak{Z}$, the Integers, can be embedded within the rationals. We will, like many others, use $\mathfrak{Q}$ as a symbol to stand for the set of Rational Numbers.

Recall that we first expanded the Whole Numbers to the Integers so that we could solve equations such as $5 + x = 3$. Solving that equation was impossible in the Whole Numbers, but becomes possible in the Integers. Now we would like to solve equations involving multiplication. We can solve $5 \cdot x = 10$ but not $3 \cdot x = 10$. It has no integer solution. We introduced *div* and *rem* to partially compensate for that, and they have their own uses as well.

It is useful to envision the number systems we are developing here as being a straight line, extending infinitely in both directions: the *Number Line*. The number 0 marks a convenient point on this line and 1 is a point somewhat to its right. The distance between 0 and 1, the unit distance, is then repeatedly marked off on the line in both directions, giving 2, 3, etc., but also –1, –2, and so on. We would now like to be able to fill in the intermediate points between the Integers.

Definitions and Properties

Our development of the Rational Numbers will be very much like our defining of the Integers. We shall define a set using properties of the Integers and an equivalence relation. The Rationals will be the set of equivalence classes. We will use the symbol, Q for the underlying set and $\mathbb{Q}$ for the set of equivalence classes in Q.

We are aiming for a system in which we can define *division* as an inverse (partial inverse) of multiplication. We would like to have a solution to $a \cdot x = 1$ or, more generally, to $a \cdot x = b$. But we won't be able to divide by zero, since we already know that $0 \cdot x$ can never be 1. So 0 has no multiplicative inverse.

Let $Q = \{ (a, b) \mid a, b \in \mathbb{Z}, \text{ and } b \neq 0 \}$. This is the basic set. Note that this is a subset of $\mathbb{Z}^2$, since the second component has a restriction.

Definition 7.1: Define a relation β on Q by: $(a, b)\ \beta\ (c, d)$ provided $a \cdot d = b \cdot c$.

Note that the definition depends only on properties of Integers.

For example (3, 10) β (9, 30) β (-6, –20), etc. Also (0, 5) β (0, 3). But nothing is equivalent to (3, 0) since zero isn't valid in the second position of our ordered pairs. (3, 0) isn't even a value here.

Theorem 7.2: The relation β is an equivalence relation in Q:

Proof: Let a, b, c, d, e, and f be in $\mathbb{Z}$. Assume b, d and f are non–zero.

(reflexive) $(a, b)\ \beta\ (a, b)$ is easy: $a \cdot b = b \cdot a$, by the commutative law of multiplication in $\mathbb{Z}$.

(symmetric) Suppose $(a, b)\ \beta\ (c, d)$. Then $a \cdot d = b \cdot c$. Applying commutativity in the Integers on both sides establishes $(c, d)\ \beta\ (a, b)$.

(transitive) Suppose $(a, b)\ \beta\ (c, d)$ and $(c, d)\ \beta\ (e, f)$. Then $a \cdot d = b \cdot c$ and $c \cdot f = d \cdot e$. Multiply both sides of the first equation by f (which is not zero) and both sides of the second by b (also not zero) and we get $a \cdot d \cdot f = b \cdot d \cdot e$. But d is not zero either, so we can cancel and get $a \cdot f = b \cdot e$ or $(a, b)\ \beta\ (e, f)$. QED

Note that this proof depends only on the laws we have established for Integers. We shall see that this will generally be the case below. We state a theorem about Rational Numbers and prove it using the rules for Integers.

Definition 7.3: The Rational Numbers, $\mathbb{Q}$, are defined to be the set of equivalence classes of the relation β on set Q.

If (a, b) is a representative of some equivalence class we will usually write it as a/b as we did above. Note that 10/3 and 30/9 are two different representations of the *same* Rational Number. We will again write the equivalence class of (x, y) for this relation as $E(x, y)$, though more precisely it is $E_\beta(x, y)$.

In particular, this means that $E_\beta(a, b) = E_\beta(c, d)$ provided $(a, b)\ \beta\ (c, d)$. This gives us a definition of equality in the Rational Numbers, of course.

The symbol β can be read "is equivalent to" because of the above theorem.

When we write "x/y is a Rational Number" we are using x/y as a shorthand for the equivalence class of the pair (x, y), not just for the pair itself.

Definition 7.4: Rational Numbers x and y are said to be *equal* if their equivalence classes are identical. We then write $x = y$.

Theorem 7.5: Equality in $\mathbb{Q}$ is an equivalence relation.

Proof: This is an immediate corollary of the above theorem. If x = E(a, b) and y = E(c, d) then x = y if and only if (a, b) β (c, d). But β is an equivalence relation. QED

Since we like to name things, in x/y we call x the *numerator*, and y the *denominator*. And of course, the denominator is never zero. But note, that since a given Rational Number has several representations, the numerator and denominator are not unique.

Definition 7.6: E(a, b) + E(c, d) is defined to be E($a \cdot d + b \cdot c$, $b \cdot d$).

So E(3, 5) + E(3, 4) = E(12 + 15, 20) = E(27, 20). Expressed in more traditional notation: 3/5 + 3/4 = 27/20.

Definition 7.7: The negative of E(a, b) is – E(a, b) = E(–a, b).

Thus, -(3/5) = -3/5. Note that (-a, b) β (a, -b).

Definition 7.8: E(a, b) – E(c, d) = E(a, b) + (–E(c, d)).

Definition 7.9: E(a, b) · E(c, d) is defined to be E($a \cdot c$, $b \cdot d$).

So E(3, 5) · E(3, 4) = E(9, 20). Again, 3/5 · 3/4 = 9/20.

Note in particular that for $a, b \in \mathbb{Z}$ it is true that: E(a, 1) + E(b, 1) = E($a + b$, 1), E(a, 1) – E(b, 1) = E($a - b$, 1), and E(a, 1) · E(b, 1) = E($a \cdot b$, 1).

Note that, very much unlike the situation with integers, where we hardly ever would use the ordered pair representation except in proofs, we normally use the ordered pair version for Rationals, though we usually write it as a/b rather than E(a, b). But when we write 3/10 = 9/30 we are just using a shorthand for E(3, 10) = E(9, 30) or even (3, 10) β (9, 30).

The Rationals have many of the properties of the Integers. Addition and Multiplication are commutative and associative. Multiplication distributes over addition and subtraction. There are a large number of theorems needed to prove this. First we need to show that addition and multiplication are well defined; i.e. independent of which pair is used to represent the operands. Then we need to prove each theorem. In the following we will often write xz in place of $x \cdot z$ to indicate multiplication, especially but not exclusively integer multiplication. Since we have used single letter "names" for numbers there should be no confusion.

Theorem 7.10: Addition is well–defined in $\mathbb{Q}$.

Proof: Let (w, x) β (y, z) and (a, b) β (c, d). Then $E(w, x) + E(a, b) = E(wb + ax, bx)$. Likewise $E(y, z) + E(c, d) = E(yd + cz, zd)$. We need to show that $(wb + ax)zd = (yd + cz)bx$. This is the same as showing $wbzd + axzd = ydbx + czbx$. But we know that $wz = xy$ and $ad = bc$. Substituting these gives the desired result. QED

Theorem7.11: Multiplication is well–defined in $\mathbb{Q}$.

Proof: Let (w, x) β (y, z) and (a, b) β (c, d). Then $(w, x)\cdot(a, b) = (wa, xb)$. Also $(y, z) \cdot (c, d) = (yc, zd)$. We need to show that $wazd = ycxb$. But $wz = xy$ and $ad = bc$. Substituting the first into the left side and the second into the right gives the result. QED

By the above theorems, we may use arbitrary representatives of rational numbers in our proofs, knowing that the results are independent of the representative chosen.

Theorem 7.12: Multiplication in $\mathbb{Q}$ is commutative.

Proof: Let a/b and c/d be Rational Numbers. Then $a/b \cdot c/d = (a{\cdot}c)/(b{\cdot}d) = (c{\cdot}a)/(d{\cdot}b)$ using properties of the Integers. But this last is $c/d \cdot a/b$. QED

We won't provide proofs of the following. They are just like the proof of the commutative law. Reduce the formula to one in which you can use the properties of Integers.

Theorem 7.13: Addition in $\mathbb{Q}$ is associative.

Theorem 7.14: Addition in $\mathbb{Q}$ is commutative.

Theorem 7.15: Multiplication in $\mathbb{Q}$ is associative.

Theorem 7.16: Multiplication in $\mathbb{Q}$ distributes over addition both on the left and right.

Theorem 7.17: The value $0 = 0/1 = E(0, 1)$ is the *additive identity* in $\mathbb{Q}$. That is: $0 + x = x$ for any Rational x.

Theorem 7.18: The value $1 = 1/1 = E(1, 1)$ is the *multiplicative identity* in $\mathbb{Q}$. That is: $1 \cdot x = x$ for any Rational x.

Theorem 7.19: For any $x = \mathrm{E}(a, b)$ in $\mathbb{Q}$, $-x = \mathrm{E}(-a, b)$ is the *additive inverse* of x.

Proof: $(a, b) + (-a, b) = (ab - ba, bb) = (0, bb)\ \beta\ (0, 1)$. QED.

Theorem 7.20: If $x = \mathrm{E}(a, b)$ is in $\mathbb{Q}$ and $x \neq 0$, then $x^{-1} = \mathrm{E}(b, a)$, has the property that $x \cdot x^{-1} = x^{-1} \cdot x = 1$. The value x^{-1} is called the *multiplicative inverse* of x. Note that $x \neq 0$ if and only if $a \neq 0$.

Some additional properties can be derived from the following equivalences

$(-a, b)\ \beta\ (a, -b)$ since $(-a) \cdot (-b) = b \cdot a$.

$(-a, -b)\ \beta\ (a, b)$ since $(-a) \cdot b = (-b) \cdot a$.

Because of this, we note that for any Rational Number a/b we can assume without loss of generality (WLOG[2]), that either a and b are both positive, or only a is negative. In other words, for any Rational we can always choose a representation for which that is true. We can also assume that a and b have no common factors (other than 1, of course). This is a consequence of the following:

Theorem 7.21: If a, b, and c are Integers with b and c non–zero, then $ac/bc\ \beta\ a/b$.

Proof: Exercise.

In particular $0/a\ \beta\ 0/1$ and $a/a\ \beta\ 1/1$.

Definition 7.22: If we choose a representation of a/b in which a and b have no common factors we say that it is expressed in *lowest terms*.

Definition 7.23: A representation of a Rational that obeys the sign convention, above, and is also in lowest terms will be called the *canonical representation* and we say the number is in *canonical form*. We normally prefer to use the canonical form for Rationals. This makes sense, since neither equality nor any of the operations depend on which representation we choose.

---

[2] Note on "WLOG". When we say that we may assume something Without Loss of Generality we mean that we are about to discuss or prove a special case, but also that the general case follows as a simple corollary of the special case. It is a fairly common technique in maths.

Theorem 7.24: If a Rational Number a/b is expressed in canonical form then the numerator and denominator are unique. In other words, if $a/b = c/d$ and both are in canonical form, then $a = c$ and $b = d$.

Proof. By contradiction. First, a and c have the same sign. Likewise b and d are both positive. If b and d aren't equal then one must be smaller than the other, so assume without loss of generality that $b < d$. Next, we know that $a \cdot d = b \cdot c$.

Therefore d divides $b \cdot c$, but, c and d have no common factors since c/d is canonical. But then d divides b by Euclid's Lemma and so $d < b$ since both are positive. The contradiction establishes the result. QED

Exercise: In the above, how can we be assured that a and c have the same sign?

Definition 7.25: We embed the Integers into the Rational Numbers using the following mapping: If x is in $\mathbb{Z}$ then its image is $E(x, 1)$ in $\mathbb{Q}$.

The following exercises can be accomplished by reducing the statements to the corresponding laws in $\mathbb{Z}$.

Exercise: Prove the substitution law for addition in $\mathbb{Q}$. If a, b, and x are rational numbers and $a = b$ then $a + x = b + x$. Let $a = a_1/a_2$, $b = b_1/b_2$, and $x = x_1/x_2$, where the subscripted values are integers.

Exercise: Prove the converse of the above, which is the cancellation law of addition.

Exercise: State and prove the substitution law of multiplication in $\mathbb{Q}$.

Exercise: State and prove the cancellation law of multiplication in $\mathbb{Q}$. Note that zero must be excluded.

Definition 7.26: We say $a/b < c/d$, where a/b and c/d obey the sign convention, provided $ad < bc$. In particular, you can use this rule if a/b and c/d are in canonical form.

Note that we define < for Rationals by reference to < for Integers since both ad and bc are Integers.

Thus $4/7 < 3/5$ and $-3/5 < -4/7$.

Exercise: Show that < is well defined and does not depend on which representation of the values is chosen.

Definition 7.27: $x \leq y$ for x, y in $\mathbb{Q}$ provided $x < y$ or $x = y$.

Theorem 7.28: The embedding of the integers into the Rational Numbers preserves less than.

Proof: exercise.

Theorem 7.29: The set $\{E(a, 1) \mid a \in \mathbb{Z}\}$ behaves like the Integers.

Proof: The function $f(a) = E(a, 1)$ is a structure preserving bijection of $\mathbb{Z}$ onto $\{E(a, 1) \mid a \in \mathbb{Z}\}$. The details are in the note, just above. The properties of sum, product, 0, 1, and ≤ are all preserved. QED

We will thus describe a Rational with a representation such as 10/5 as an Integer, relying on this injection, since 10/5 is equivalent to 2/1.

The following theorems concerning ≤ and < are all easy to prove.

Theorem 7.30: Relation < transitive in $\mathbb{Q}$.

Theorem 7.31: Relation < obeys the law of trichotomy in $\mathbb{Q}$.

Theorem 7.32: Relation ≤ is reflexive in $\mathbb{Q}$.

Theorem 7.33: Relation ≤ is transitive in $\mathbb{Q}$.

Theorem 7.34: Relation ≤ is antisymmetric in $\mathbb{Q}$.

Theorem 7.35: Relation ≤ is *total* in $\mathbb{Q}$. If x and y are in $\mathbb{Q}$ then $x \leq y$ or $y \leq x$.

Theorem 7.36: If x and y are in $\mathbb{Q}$ and $x \leq y$ then $x + z \leq y + z$ for any z in $\mathbb{Q}$.

Theorem 7.37: If x and y are in $\mathbb{Q}$ and $0 \leq x$ and $0 \leq y$, the $0 \leq x \cdot y$.

Corollary 7.38: If x, y *and* z are in $\mathbb{Q}$ and $0 \leq x$ and $y \leq z$ then $xy \leq xz$.

Notably missing from this list is a theorem that says that a bounded non-empty set of Rational Numbers has a least element. In fact, it is false. We will explore that in a bit. In particular, $\{x \mid x \text{ is a Rational Number and } x > 0\}$ has no least element.

Definition 7.39: Let w/x and y/z be in $\mathbb{Q}$ with $y \neq 0$. Then the *quotient* of w/x and y/z, denoted by $(w/x) \div (y/z)$, is $(w/x) \cdot (z/y) = (w{\cdot}z)/(x{\cdot}y)$. The operator $\div$ is called *division*.

We also use / to denote division. Thus $(w/x) \div (y/z)$ can be written, instead, as (w/x) / (y/z). Note that / is technically used in two different ways here, as the separator of the numerator and denominator of a Rational and as an operator. However, we may also consider the former use to be division also for the following reason:

Let $w/1$ and $x/1$ be integer values in $\mathbb{Q}$. Then $(w/1) \div (x/1) = (w{\cdot}1)/(1{\cdot}x) = w/x$. So the Integer w, considered as a Rational, and x, also considered as Rational, when divided, yield the Rational w/x. Thus w/x, though just a notation, is consistent with Rational division.

Now, the equation $3 \cdot x = 10$ has a solution, namely 10/3.

But we can say more than that.

Theorem 7.40: Let $r \cdot x = s$ be an equation with variable x. Let r and s be in $\mathbb{Q}$ with $r \neq 0$. Then $x = s/r$ (the quotient) is the unique solution to the equation. The value s/r is Rational since $r \neq 0$.

Proof: $r \cdot (s/r) = s$ as can be easily shown. To show that this solution is unique, let a/b be a solution to $r \cdot x = s$. Then $r \cdot (a/b) = s$. If we multiply both sides by b we see that $ra = sb$. Therefore (s, r) β (a, b). QED

Thus we can solve simple equations with Rational (not just integer) coefficients.

For example $(3/5) \cdot x = 7/8$ has solution $(7/8)$ / $(3/5) = (7/8) \cdot (5/3) = 35/24$.

In fact the equation $p \cdot x + q = r$ is easily solvable using the tools developed above, assuming p, q, and r are Rational with $p \neq 0$. The solution is $(r - q)/p$.

Denseness and Distance in the Rational Numbers

The following theorem is what makes the Rational Numbers "interesting."

Theorem 7.41: Between any pair of distinct Rational Numbers there is a third, distinct from the two given numbers.

Proof: Let $r < s$ be Rational, where $r = w/x$ and $s = y/z$ are in canonical form. Let $t = (r + s)/2 = (wz + xy)/(2xz)$. This number is called the *midpoint* or the *average* of r and s. Note that it obeys the sign convention. We now show $r < t$ and $t < s$. We know that $wz < xy$, $x > 0$, $z > 0$ (see the monotonic property of multiplication in the integers mentioned on pg. 71). Therefore $xwz < xxy$ and $zwz < zxy$. But then: $r < t$ since $w \cdot (2xz) = 2wzx = xwz + xwz < xwz + zxy = x \cdot (wz + xy)$.

Similarly $t < s$ since $z \cdot (wz + xy) = zwz + zxy < xyz + zyz = 2xyz$. QED

However, between r and t there is another and another and another.

Definition 7.42: We say that the Rational Numbers are *dense* since between any two Rationals there is a distinct Rational. The integers, by way of contrast are called *discrete* or *sparse*. Between 3 and 4 there is no integer but there are an infinity of Rationals, of course.

So, while the Integers are infinite, the Rationals seem to be infinite in a different way: infinite everywhere, so to speak. As a consequence, if you have a Rational Number r there is no concept of a *next* Rational following r. If you choose any Rational close by, another will be closer still. If you need numbers for measurement, then, Rational Numbers are what you want. You can measure any length to as fine an accuracy as desired.

Definition 7.43. The *absolute value* of a Rational Number r is $|r|$ defined to be r if $r \geq 0$ and $-r$ otherwise.

Definition 7.44: The *distance* between two Rational Numbers can be defined as the absolute value of their difference. $d(r, s) = |r - s|$. The distance between Rationals can be as small as you like. This is because the average of two Rationals, $(r + s)/2$, is closer to r and to s than r and s are to each other. This will be very important later.

Cardinality of the Rational Numbers

So, the Rationals have a very interesting property that isn't shared with the Integers. But they do share one interesting property. In a mathematical sense, there are just as many Integers as there are Rationals. It seems like there must be more, but there aren't. Before we can justify that, however, we need to define what we mean by "more".

Clearly, if I have 3 sheep and you have 5, then you have more. We know that $3 < 5$, but there is another reason. If you and I don't know about numbers at all, we can still determine that you have more, just by matching sheep. We each separate out one sheep from our herds and consider the remainder. We do it again, until I don't have any more to match, but you still do. Expressed in math terms, we were able to define a bijection between my herd and a proper subset of yours: a subset that wasn't the whole set. But we were not able to define a bijection between my herd and all of yours.

Definition 7.45: We say that a set A has a *cardinality less than or equal* to that of a set B, provided that there is an injection from A into B.

Definition 7.46: We say that sets A and B have the *same cardinality* provided that the cardinality of each is less than or equal to that of the other.

Cardinality is just a nice big word for size, though now it has a precise meaning. *Size* can be interpreted differently, of course. Suppose my sheep are a lot larger than yours, for example.

By this definition, any set of five things (say sheep) has the same size as any other set of five things (say coins). But it also means that the cardinality of the even numbers is the same as the cardinality of the integers. And this is true even though the even numbers are a proper subset of the integers. Still, we can define a bijection from the even Integers to the Integers, and it is quite simple. Let $f(x) = x/2$ for any even number x. Then the domain of f is the even numbers, but the range is all of the integers. We can see the matching as follows:

… –8 –6 –4 –2 0 2 4 6 8 … (all of the even numbers)

… –4 –3 –2 –1 0 1 2 3 4 … (all of the integers)

The inverse function is g(x) = 2x, of course.

In fact, the definition of *infinite* is that an infinite set can be put into bijection with (matched with) a proper subset of itself.

But we can also match the Integers with the Rationals. Note that, as a consequence, the set of subsets of the Integers, $2^{\mathbb{Z}}$, has the same cardinality as the set of subsets of the Rationals, $2^{\mathbb{Q}}$.

Theorem 7.47: The cardinality of the positive Rational Numbers is the same as the cardinality of the Natural Numbers.

To prove this, imagine an infinitely large table. The rows are labeled with the Natural Numbers and so are the columns. The intersection of a row and a column is some particular representation of a positive Rational. Each Rational is represented several times in this table of course. We show a small portion of the table below. In this table we will follow (minor) diagonals from Northeast to Southwest, passing through each element, counting as we go. Counting, of course is matching Natural Numbers with the things counted. We start at the diagonal in the Northwest corner. We first encounter (1, 1) which is the Rational 1 and so we count it. Since it is first, it gets number 1. Then we move on to the next diagonal and see (1,2), or 1/2, and count it as 2. Then (2, 1), or 2, which gets counted as 3. Next (1, 3) is counted as 4. Next, however, we encounter (2, 2) which is the Rational 1 again so we just skip it. Then (3, 1) is the fifth element actually counted. We continue this way, skipping any Rational that we have already seen in the process.

	1	2	3	4	5	6	7	8	9
1	(1,1)	(1,2)	(1,3)	(1,4)	(1,5)	(1,6)	(1,7)	(1,8)	(1,9)
2	(2,1)	(2,2)	(2,3)	(2,4)	(2,5)	(2,6)	(2,7)	(2,8)	(2,9)
3	(3,1)	(3,2)	(3,3)	(3,4)	(3,5)	(3,6)	(3,7)	(3,8)	(3,9)
4	(4,1)	(4,2)	(4,3)	(4,4)	(4,5)	(4,6)	(4,7)	(4,8)	(4,9)
5	(5,1)	(5,2)	(5,3)	(5,4)	(5,5)	(5,6)	(5,7)	(5,8)	(5,9)
6	(6,1)	(6,2)	(6,3)	(6,4)	(6,5)	(6,6)	(6,7)	(6,8)	(6,9)
7	(7,1)	(7,2)	(7,3)	(7,4)	(7,5)	(7,6)	(7,7)	(7,8)	(7,9)
8	(8,1)	(8,2)	(8,3)	(8,4)	(8,5)	(8,6)	(8,7)	(8,8)	(8,9)
9	(9,1)	(9,2)	(9,3)	(9,4)	(9,5)	(9,6)	(9,7)	(9,8)	(9,9)

We are defining a function from the Natural Numbers onto the positive Rationals since the table contains all of the positive Rationals. Since we count each rational only once, it is a bijection. Note that in deciding whether to count a representation as a Rational or not we need only make a finite number of decisions, comparing it to the finite number of Rationals we have already counted. It is a bit messier, but fundamentally the same to map all of the integers onto all of the Rationals. The cardinality of the Integers (or the Natural Numbers, for that matter) is called *Aleph Nought*: $\aleph_0$. It is also called *Aleph Null*. Aleph is the first letter of the Hebrew alphabet. Aleph Nought is the name for the cardinality of any set that can be matched with the Natural Numbers. Such sets are called *countably infinite*. If a set is finite or countably infinite it is called *countable*. So, the Rational Numbers are countable.

Decimals and the Rational Numbers

A decimal number of finite length (or equivalently one that ends in all zeros) represents a Rational Number. For example, 2.33 means 2 + 33/100 = 233/100. You learned a long division process in school to take such things in the other direction. However, a decimal number that is infinite but repeating on the right is also a Rational Number.

For example $x = 2.5323232...$ repeating "32" forever is Rational.

This is normally indicated by overlining the repeating segment, rather than repeating it. So $x = 2.5\overline{32}$ and then

$100 \cdot x = 253.232323...$ repeating. If we subtract x from $100{\cdot}x$ we get

$99x = 250.7$. Multiply both sides by 10 to get $990x = 2507$, so $x = 2507/990$.

In general, multiply x by 10^n, where n is the length of the repeating segment. Here it was just 2. Then the subtraction eliminates the repeating part.

Here is another interesting fact. The repeating decimal $z = 0.9999...$ repeating 9's forever is equal to 1. The above technique will demonstrate it. You can also see this with some intuition. How close is z to 1? If we truncate it anywhere it is pretty close. Extending it with another 9 makes it closer. But the distance between the truncated version and 1 decreases by 90% each time we extend with a 9. We will see Zeno's Paradox soon, which uses a similar idea.

I suggest you play with the above for something like $x = 0.\overline{33}$ and $y = 0.\overline{66}$. Note that the second 3 and 6 here are superfluous. The format $x = 0.\overline{3}$ works just as well. The 3 repeats forever.

Also try $z = 0.\overline{142857}$ repeating a six digit segment forever. Explore the decimal expansion of 1/17 and show that it repeats.

Exercise: Use Theorem 6.61 to prove that the decimal expansion of any Rational Number repeats infinitely, though perhaps with all zeros.

Summary So Far

The Integers are good for counting things. The Rationals are better, however, for measuring things. We can distinguish 6 5/8 (or 6 + 5/8) from 6 3/8, for example, measuring the length of your hand and that of your sister, perhaps. The Rationals retain most of the desirable properties we developed for the integers, except, notably, a guaranteed minimum value in a non-empty set of Rational Numbers that is bounded below.

The notion of an equivalence relation came in handy again. It let us give a concise definition to the Rationals and their operations and relations, but it also let us reduce the proofs of the important theorems to known facts about the Integers. Multiplication is commutative in the Rationals because of the properties of the Integers.

We saw the, perhaps surprising, result that the Rationals have the same cardinality as the Integers, though the Integers are a tiny subset of the Rationals. There are an infinite number of Rationals for every Integer, but the cardinality is the same. Infinity is odd. We will see that it is odder yet.

We also saw that we could solve more equations in the Rationals than in the Integers, but not all. We need to extend yet again to be able to solve a larger class of equations.

Earlier we noted the Abstract Algebra notions of Group and Ring. Another very important structure is a *Field*. Roughly speaking, a Field is a Ring in which multiplication is commutative and every non zero element has a multiplicative inverse. We have seen above that the Rational Numbers form a Field. But the theory of Fields in general is itself very rich and goes far beyond the Rationals. In fact, the Integers modulo 7 (or any prime number) forms a Field also. If the base isn't prime, such as 12, then you still get a Ring, but you won't have a multiplicative inverse for every element. For example, in the Integers mod 7, if you multiply 5 by 3 you get 1 (mod 7) so 5 and 3 are multiplicative inverses of each other.

In ancient Greece, the Pythagoreans were very obsessed with integers and Rational Numbers. They associated the number 1 with divinity. They seem to have thought that everything in the world was comparable to everything else by ratios of integers. But then they discovered that a square whose side is 1 has a diagonal that cannot be Rational. This caused them a large amount of grief. Suddenly they find that $x^2 = 2$ has no Rational solution. They really want to measure that diagonal, but they need a better way to measure in order to achieve it. So can't be a *Rational Number*. What sort of number could it be?

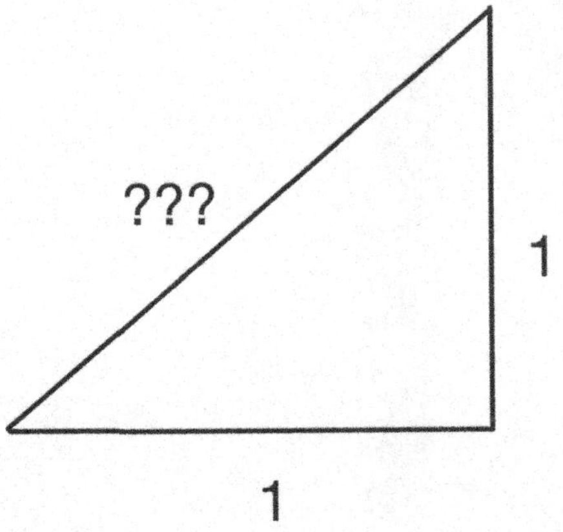

Not Rational

Zeno's Daughter

Hitting the Mark Since 450 BCE

8 Infinite Sequences

We will need to understand some things about infinite sequences in order to extend the Rational Numbers and define the Real Numbers.

Definition and Properties

Definition 8.1: An *infinite sequence* is a function from some infinite subset of $\mathfrak{N}$ to some other set. The domain set is usually $\mathfrak{N}$ itself, but need not be. The range set is often some set of numbers, but that isn't essential. Normally, for a sequence function x, we write x_n instead of $x(n)$ for a value n in the domain. Sometimes we just list a few values in order of the domain element that produced them. So the function $y(n) = y_n = n$, the identity function, might be given as just: 1, 2, 3, 4, …. Likewise, $z(n) = z_n = 1/n$ is 1, 1/2, 1/3, 1/4, ….

In the following, we assume that "sequence" means infinite sequence.

The range of a sequence can be finite or infinite. For example, a constant sequence, such as 2, 2, 2, 2, … only has a single element in the range, whereas the identity function yields an infinite range.

Let x be a sequence. We might use any of the following notations to indicate x. Note the similarity to set notation, though the intent is different. The "order" of elements in such a representation is important.

$x = x_1, x_2, x_3, \ldots = {}_{n=1}\{x_n\}^{\infty} = \left\{x_n\right\}_{n=1}^{\infty} = \{x_n \mid n = 1, 2, 3, \ldots\} = \{x_n \mid n = 1, 2, 3, \ldots, \infty\} = \{x_1, x_2, x_3, \ldots\} = \{x_n \mid x \in \mathfrak{N}\} = \{x_n\}$

Many times we can utilize a formula to give the elements of a sequence, such as: $x_n = 1/2^n$ for the sequence 1/2, 1/4, 1/8, 1/16, 1/32, … We would then write $x = \{1/2^n \mid n \in \mathfrak{N}\}$ or some variation.

The examples here use Rational Numbers for the range values, but that isn't essential to the concept. The values can be Rational or drawn from other systems. We will use sequences of Rational Numbers to define the Real Numbers (reals), but then can also have sequences of reals. But if you have two jellybeans you can have an infinite sequence of jellybeans by defining j(n) = "the red jellybean" if n is even and j(n) = "the blue jellybean" if n is odd. You don't need an infinite number of jellybeans. So a constant function, such as we mentioned above doesn't imply that there are an infinite number of 2's. It is the domain of the function that is infinite, not necessarily the range.

We will assume for now that the range set of a sequence is some subset of the Rational Numbers.

In defining a sequence we often start with 1 and use all of $\mathfrak{N}$ for the domain, but that is not essential. When we deviate, however, it is useful to be explicit. We can also use this idea to define *subsequences* of a given sequence. Occasionally we even start with 0, resulting in a function from $\mathfrak{W}$ to $\mathfrak{Q}$ or to some other set.

For example, let $x = \{x_n \mid n = 1, 2, 3, \ldots\}$ be an arbitrary sequence. Then $y = \{x_n \mid n \in \mathfrak{N}\ n > 5\}$, or $z = \{x_n \mid n \in \mathfrak{N}, n \text{ is even}\}$ are two subsequences. The elements of a subsequence are a selection of the elements of the original, with the correspondence between Natural Numbers and range values preserved, or at least the order of elements preserved.

Given a sequence x and a value of N, the values of the (finite) subsequence x_n for $n \leq$ N is called a *prefix* of the sequence. The remainder of the sequence is sometimes called the *tail*. Technically, a prefix isn't a sequence, since it is finite.

Here and in the future some symbols will be useful as shorthand. The symbol $\forall$ is often used to mean "for every" and it is followed by a description of some value. For example $\forall\ x \in \mathfrak{N}$ is a way of naming a variable x and making statements about all such values as long as they are Natural Numbers. It is just shorthand for "for every Natural Number, x"

Another important symbol is $\exists$, short hand for "there exists". The first Peano Axiom might be stated as "$\exists$ Natural Number named 1."

The symbols $\forall$ and $\exists$ are called *quantifiers* and they are used throughout math and logic.

Definition 8.2: An infinite sequence, $x = \{x_n\}$, is called *bounded above* provided that $\exists$ Natural Number B such that $\forall\ n \in \mathfrak{N}$ it is true that $x_n <$ B.

Definition 8.3: An infinite sequence, $x = \{x_n\}$, is called *bounded below* provided that $\exists$ Natural Number B such that $\forall\ n \in \mathfrak{N}$ it is true that $x_n > B$.

Definition 8.4: An infinite sequence, $x = \{x_n\}$, is called *bounded* provided that $\exists$ Natural Number B such that $\forall\ n \in \mathfrak{N}$ it is true that $|x_n| < B$.

The example sequences above have all been bounded below and some, but not all, have been bounded. A sequence that is not bounded above or below is: $1, -2, 3, -4, \ldots$

Definition 8.5: An infinite sequence, x, is called *monotone increasing* provided that $\forall\ n \in \mathfrak{N}$ it is true that $x_n \leq x_{n+1}$.

Definition 8.6: An infinite sequence, x, is called *monotone decreasing* provided that $\forall\ n \in \mathfrak{N}$ it is true that $x_n \geq x_{n+1}$.

Definition 8.7: An infinite sequence, x, is called *monotone* provided that it is either monotone increasing or monotone decreasing.

Note that a constant sequence is both monotone increasing and monotone decreasing by this definition. If we change the $\leq$ and $\geq$ relations to $<$ and $>$, respectively, we have strictly monotone increasing or *monotone strictly increasing* (as you prefer) and similarly for decreasing.

The sequence $1, -1, 1, -1, \ldots$ is neither increasing nor decreasing, though it is bounded.

Traditionally small positive numbers have been named with lower case Greek letters such as epsilon, ε, and delta, ∂. Also, s. t. is shorthand for "such that."

Definition 8.8: We say that a sequence of Rationals $\{x_n\}$ *converges to 0*, provided that $\forall$ Rational $\varepsilon > 0\ \exists\ N \geq 1$ s.t. $\forall\ k > N$ it is true that $|x_k| < \varepsilon$. We also say the sequence has *limit* 0.

This is just a way of saying that the values of the sequence get small eventually and then stay small. How small? As small as you please. For every ε if we go out far enough (beyond the corresponding N) the values are smaller than ε. Notice that the N in this definition depends on ε. Generally speaking, the smaller the ε, the larger the required N.

There is nothing essential about the numbers being Rational, here, but so far we don't have any larger class of numbers, so we will stick to Rationals for the time being.

The construction "∀... ∃ ... such that..." that we have seen here is extremely important and extremely common in math. It is used pervasively in Calculus, for example. One of the harder concepts to grasp, for many students, is the idea of a limit. The derivative and the integral both depend on a solid understanding of limits. We have only used it with sequences, but it has much larger applicability. But understanding the concept is critical to your development as a mathematician. To show that a given sequence converges to a particular number, such as 0, we must show that no matter how small an ε we choose it will be possible to find some N for which every value in the sequence beyond the Nth is closer to the given value (here 0) than ε. We can't have a fixed N, like 10,000, as it depends on ε. But we must be able to do this for every small positive ε. Note that we don't need to consider large values of ε, since if it works for a small value the same value of N works automatically for all larger values of ε. We can *squeeze* the sequence arbitrarily beyond a given N.

Example: $\{1/2^n\} = \{1/2, 1/4, 1/8, 1/16, 1/32, \ldots\}$ converges to 0. Choose any $\varepsilon > 0$. Let N be such that $1/2^N < \varepsilon$. Then if $k > N$, $|1/2^k| < 1/2^N < \varepsilon$.

Example: $1/2, 1/3, 1/4, 1/5, \ldots = \{1/n \mid n = 2, 3, \ldots\}$ converges to 0.

Exercise: Show that the above example converges to 0.

Example: 1/2, 2/3, 3/4, 4/5, 5/6, ... does not converge to 0.

Example: 1, –1, 1, –1, ... does not converge to 0.

We can define various operations on sequences of numbers. Again, we assume they are Rational for now.

Definition 8.9: If $x = \{x_n\}$ is a sequence and r is a Rational Number then $r \cdot x = x \cdot r = \{r \cdot x_n \mid n \in \mathfrak{N}\}$. In other words, multiply each element by r to get a new sequence.

Exercise: If $x = \{x_n\}$ is a sequence of Rationals that converges to 0, and r is a Rational Number, then $r \cdot x$ also converges to 0.

Definition 8.10: Let $c = \{ x_n \mid n \in \mathfrak{N}\}$ and $d = \{ y_n \mid n \in \mathfrak{N}\}$ be sequences. Define the negative, sum, difference, product, and quotient of c and d as follows.

$-c = \{- x_n \mid n \in \mathfrak{N}\}$.

$c + d = \{x_n + y_n \mid n \in \mathfrak{N}\}$.

$c - d = \{ x_n - y_n \mid n \in \mathfrak{N}\}$.

$c \cdot d = \{ x_n \cdot y_n \mid n \in \mathfrak{N}\}$.

If d satisfies $\exists$ N ≥ 1 s.t. $y_n \neq 0$ $\forall n > N$, then we define

$c/d = \{ x_k / y_k \mid k = N + 1, N + 2, N + 3, \ldots \}$.

For the quotient we simply omit the earlier, prefix, section of both sequences in which d potentially has zero values. That is enough for what we want to do, since we are really interested only in what goes on in most sequences for large values of the subscript.

Exercise: Any tail of a sequence that converges to 0 also converges to 0.

Note that the rule for multiplication by a constant given above has the same result as multiplication by a constant sequence.

Definition 8.11: If $c = \{ x_n \mid n \in \mathfrak{N}\}$ is a sequence, define $|c| = \{ |x_n| \mid n \in \mathfrak{N}\}$.

Definition 8.12: Let $c = \{ x_n \mid n \in \mathfrak{N}\}$ and $d = \{ y_n \mid n \in \mathfrak{N}\}$ be sequences. Define the distance between c and d to be $|c - d|$. This is, $\{ |x_n - y_n| \mid n \in \mathfrak{N}\}$, of course.

Note that Definitions 8.11 and 8.12 define sequences, not numbers.

It should also be clear from the definitions that the sum and product of sequences obey the rules we have come to expect. Addition is commutative and associative. Multiplication is commutative and associative. The sequence of all zeros acts like an additive identity. The sequence of all ones acts like a multiplicative identity. Multiplication distributes over addition and subtraction, both on the right and left sides. Additive inverses are easy. Multiplicative inverses are not so easy, since a sequence might include zeros. But if a sequence includes no zeros then the sequence of multiplicative inverses of the individual elements is a multiplicative inverse of the original. Likewise if a sequence has no zeros past some given index, then, if we are willing to omit prefixes of sequences, we can define a multiplicative inverse for the tail of the original.

Definition 8.13: A *maximum element* of a sequence $x = \{x_n\}$ is an element x_k of x such that $x_n \leq x_k$ for every n. If there are several such elements, we take the one with the lowest subscript to be *The* Maximum Element. Denote the maximum element of sequence x by $\max(x)$ provided it exists.

Note that not every sequence has a maximum element. Constant sequences do as do monotone decreasing sequences. A monotone strictly increasing sequence does not, however.

Examples: 1, 2, 3, … has no maximum element, nor does any sequence that contains this as a subsequence. The sequence 1, 2, 1, 2, 1, … has (the first) 2 for its maximum element. In a monotone decreasing sequence, the first element is the maximum element.

Lemma 8.14: Suppose x is a sequence and y is a subsequence containing all but a finite number of elements of x. Suppose $\max(y)$ exists. Then x also has a maximum element and the $\max(x) \geq \max(y)$.

Proof: Sequence x has only finitely many elements not in y. If any of these are larger than or equal to $\max(y)$ then the first occurrence of the largest of them is $\max(x)$. Otherwise $\max(y)$ is also $\max(x)$. QED

The following theorem will be important in the next chapter.

Theorem 8.15: Every sequence contains a monotone subsequence.

Proof: Let $x = x_1, x_2, x_3, \ldots$ be an arbitrary sequence and consider the sequence of subsequences X_n which is x but omitting the first n elements:

$X_1 = x_2, x_3, \ldots = \{x_n \mid n > 1\}$
$X_2 = x_3, x_4, \ldots = \{x_n \mid n > 2\}$
$X_3 = x_4, x_5, \ldots = \{x_n \mid n > 3\}$
$X_4 \ldots$, etc.

Note, in particular, that each of these is also a subsequence of the previous one, containing all but one element.

There are two cases to consider. (a) suppose every X_k has a Maximum Element. Note that if $m > n$ then the $\max(X_m) \le \max(X_n)$, but then these maximum elements form a monotone decreasing sequence in x.

(*b*) Suppose that some X_k fails to have a Maximum Element. Then X_{k+1} also fails to have a maximum element or its maximum element would also be the maximum element of X_k. By induction, none of the X sequences beyond k have a maximum element. In particular if $n > k$ then there is an element $x_m \ge x_n$ where $m > n$ since x_n isn't a maximum element. The same reasoning gives another, still larger element, beyond x_m, etc. This results in a monotone increasing sequence. QED

The above proof has an interesting feature. There are two cases to consider, but the proof gives no hint about which case applies to a given sequence. Thus it assures us that the sequence has a monotone subsequence but gives no help whatever in finding it. It would be more satisfying to have a *constructive proof*, which produced a subsequence from an actual example. But this *existence proof* has a different, less satisfying nature. Imagine, for example, that you have a machine that can generate random numbers one after the other. Consider a sequence of numbers generated by this machine. Does such a sequence exist? If it generates numbers one after the other as stated, then no, only a prefix exists at any given moment. If you can further imagine a machine that generates them "all at once" then yes, the sequence exists and has a monotone subsequence. But you can't discover what that subsequence is in any effective way by examining prefixes of the whole. I have trouble imagining the latter machine, by the way. On the other hand, I have no real trouble with the statement "Consider an infinite sequence of random numbers." I'm happy to discuss its properties, even though it is imaginary in some sense. Infinity is tricky.

Some mathematicians will reject such "non-constructive" proofs. There is a subfield of maths called Constructive Mathematics(+) that requires that we be more explicit. It is a form of constraint on their thinking that can lead to creativity in the explorations of those who practice it. What can we take to be true (i.e. derivable from the axioms) if we reject existence proofs that don't produce a particular value? The discipline is difficult, but no less valuable.

Infinite Series

Here are a few ideas about the related topic of *infinite series*. An infinite series is an infinitely long sum. For example 1+1+1+1+... Of course, if you add a lot of 1's together the sum gets larger and larger without bound, so it makes sense to say that such a sum is infinite or that it *diverges to infinity*. But are all such sums infinite?

To motivate the idea, let's take a little excursion to ancient Greece. Well, it is now in Italy, but then it was Greek. The philosopher Zeno (Zeno of Elea who lived about 500 BCE) stated a famous paradox that is said to prove that motion is impossible, or at least just an illusion. The argument goes like this.

If you try to shoot an arrow at a target it can never reach it! Before it can reach the target it must first go half way. But from there it must again go half the remaining distance. And from there half of what remains. But this process never ends and so the arrow cannot reach the target. Therefore motion is just a delusion.

A more modern interpretation would say that that the arrow must traverse an infinite number of distances each half of the previous. If the original distance to the target is 1, then the distances to be added are 1/2, 1/4, 1/8, 1/16, 1/32, ... The sum is then

1/2 + 1/4 + 1/8 + 1/16 + 1/32 + ... Written in modern symbols this is the sum of an infinite set: $\sum(1/2^n \mid n \in \mathfrak{N})$ or $\sum_{n=1}^{\infty} 1/2^n$ While no one outside Zeno's school of thought believes that motion is impossible, the paradox illustrates that a mathematician should consider that sum to be equal to 1. But there is a bit of math yet to learn to put that idea on a firm footing.

In order to define what we mean by something like $\sum(x_n \mid n \in \mathfrak{N})$, where $\{x_n\}$ is a sequence, we first consider a sequence $S = \{S_n\}$ of finite sums.

$S_1 = x_1$, $S_2 = x_1 + x_2$, $S_3 = x_1 + x_2 + x_3, \ldots, S_k = \sum_{i=1}^{k} x_i$, ... etc.

These sums are just the finite prefixes of the original sequence. We call them *partial sums*. We ask what happens to these partial sums. For Zeno's sequence the partial sums are 1/2, 3/4, 7/8, 15/16, 31/32, ... which certainly seems to be getting closer to 1. As close as you like, actually. And none of the partial sums are greater than 1. We don't yet have a satisfying way to define "convergence to 1" but will have soon enough. The problem is that the Rational Numbers can be shown to have a lot of "holes." This means that something that seems to "converge" might be getting closer to something that isn't Rational. The value 1 is Rational, of course, so we could give a restricted definition, but choose to wait until we can do better.

Note, of course, that for an infinite series to converge to something less than infinite, it is essential that the underlying sequence of terms converge to zero though we haven't proven that theorem here. But that is not sufficient to guarantee convergence. For example the so called harmonic sequence $\{1/n \mid n \in \mathfrak{N}\}$ does converge to zero, but can be shown to have an infinite sum. To see this, note that the second term is ½. The sum of the next two (1/3 + 1/4) is more than ½. Then the sum of the next 4 is also more than ½, etc. Thus $\sum\{1/n \mid n \in \mathfrak{N}\}$ must grow to infinity. In other words, the sequence of partial sums increases without bound. Very interesting. On the other hand, the alternating harmonic series $\sum\{(-1)^{n+1}/n \mid n \in \mathfrak{N}\}$ converges to ln(2), though we won't show the proof. You can explore the *alternating series test*(+) (as well as logarithms) to see why it converges.

We examined the geometric sum earlier (pg. 60) and discovered:

$\sum\{r^n \mid 0 \le n \le k\} = (1 - r^{k+1})/(1 - r)$.

Now, if $|r| < 1$ and if we consider this as the sequence of partial sums as k increases we get (since r^{k+1} converges to zero when $|r| < 1$:

$\sum\{r^n \mid n \in \mathfrak{W}\} = 1/(1 - r)$ if $-1 < r < 1$.

Now, if we consider r to be a variable, rather than a fixed real number, this defines a function when $|r| < 1$. This sort of idea can lead to defining functions as infinite series, which is an important topic in Calculus.

3.14159265358979323846264338327950288419716939937510582097494459230781640
6286208998628034825342117067982148086513282306647093844609550582231725359
4081284811174502841027019385211055596446229489549303819644288109756659334
4612847564823378678316527120190914564856692346034861045432664821339360726
0249141273724587006606315588174881520920962829254091715364367892590360011
3305305488204665213841469519415116094330572703657595919530921861173819326
1179310511854807446237996274956735188575272489122793818301194912983367336
2440656643086021394946395224737190702179860943702770539217176293176752384
6748184676694051320005681271452635608277857713427577896091736371787214684
4090122495343014654958537105079227968925892354201995611212902196086403441
8159813629774771309960518707211349999998372978049951059731732816096318595
0244594553469083026425223082533446850352619311881710100031378387528865875
3320838142061717766914730359825349042875546873115956286388235378759375195
7781857780532171226806613001927876611195909216420198938095257201065485863
2788659361533818279682303019520353018529689957736225994138912497217752834
7913151557485724245415069595082953311686172785588907509838175463746493931
9255060400927701671139009848824012858361603563707660104710181942955596198
9467678374494482553797747268471040475346462080466842590694912933136770289
8915210475216205696602405803815019351125338243003558764024749647326391419
9272604269922796782354781636009341721641219924586315030286182974555706749
8385054945885869269956909272107975093029553211653449872027559602364806654
9911988183479775356636980742654252786255181841757467289097777279380008164
7060016145249192173217214772350141441973568548161361157352552133475741849
4684385233239073941433345477624168625189835694855620992192221842725502542
5688767179049460165346680498862723279178608578438382796797668145410095388
3786360950680064225125205117392984896084128488626945604241965285022210661
1863067442786220391949450471237137869609563643719172874677646575739624138
9086...

= 4/1 – 4/3 + 4/5 – 4/7 + 4/9 – 4/11 + 4/13 – 4/15 + ...

9 The Real Numbers

Just as we motivated the earlier extensions by the inability to solve certain equations, we shall do the same here. Consider the equation $x^2 = 2$. This isn't like the equations we considered earlier, since the variable is multiplied by itself, not just by a constant. But this equation has no Rational solution.

Suppose it did. Suppose that a Rational, r/s, in canonical form, had the property that $(r/s)\cdot(r/s) = 2$. We can simplify this to $r\cdot r = 2\cdot s\cdot s$. Now it is clear that two divides the right hand side of this, so 2 must also divide the left side. But Theorem 6.68 (Euclid's Lemma) says that, then, 2 must divide r. So r is even and $r = 2t$ for some integer t. But then the left side is $4\cdot t\cdot t$ and then $4\cdot t\cdot t = 2\cdot s\cdot s$. But then $2\cdot t\cdot t = s\cdot s$. So the same reasoning as above says that s must be even. But r/s was originally in canonical form, so we have a contradiction.

Exercise: If k is any positive prime number then the solution to $x^2 = k$ is also not Rational. In fact, the same is true provided that k is not, itself a perfect square like 16 or 25. The proof is only a bit more complex.

However, even though there is no Rational whose square is 2 you can find a Rational whose square is as close to 2 as you like. In fact you can find a Rational whose square is as close as you like, but larger than 2, and another as close as you like, but smaller. We can *squeeze* the desired value between Rationals. This is the power of denseness of the Rational Numbers. The proof goes like this.

We start with two Rational values, s, and t. We want $s^2 < 2$ and $t^2 > 2$, but otherwise it doesn't matter what they are. Suppose we let $s = 1$ and $t = 3/2$. We will also define two sequences in which to remember the values we compute. Let $L_0 = s = 1$, and $U_0 = t = 3/2$. It is easy to check that this s and t have the desired properties. We will now take the average of these two numbers $(s + t)/2$ and get 5/4. If we square this we get 25/16 which is a bit less than 2, so we replace s with this value. We also let $L_1 = 5/4$, but $U_1 = U_0 = 3/2$, remains the old value of t. Now we have $s = 5/4$ and $t = 3/2$. Note that the distance between s and t was originally ½, but now it is ¼. If we repeat the process with the new values, averaging s and t, we get 11/8. The square is 121/64, still less than 2 so the new $L_2 = s$ is 11/8 (With $U_2 = U_1$) and the two values are now 1/8 apart, half the previous distance. U_2 is set to U_1 since we don't yet have a new upper estimate. If we repeat again we get 23/16, whose square is 529/256 and this time the square is bigger than 2, so 23/16 replaces $t = U_3$. We now have $L_3 = s = 11/8$ and $U_3 = t = 23/16$, which differ by 1/16. So our s is closing in from below and t from above, though only one will change at each step. Moreover the values are getting closer and closer together *without limit*. We can make the difference as small as we like, but one value is a bit small and the other a bit large. Likewise the L sequence gets closer and closer from below and the U sequence closer and closer from above. Also note that the values in the L sequence get closer and closer to *each other*. The same is true of the U sequence. I'll call this process of defining two sequences by taking averages the *Squeeze Process*, but it might also be called the *Cauchy Process*. We are about to learn why.

So, while we called the Rationals "dense" it seems that they leave some holes. There is a missing value: the one we call the *square root of* 2 or $\sqrt{2}$ We know it isn't Rational, but it is as close as you like to many Rationals. If we want to think of it as a number then we need to extend yet again. There are a few traditional ways to do this. The one we shall see follows up on the squeeze process of the previous paragraph: a sequence of numbers whose elements get closer together without bound. These are called *Cauchy sequences* after Baron Augustin-Louise Cauchy (1789 - 1857). The basic idea of that approach is to define a set consisting of all such Cauchy sequences of Rational Numbers and then define an equivalence relation on them. The real numbers are then defined to be the equivalence classes. The course I originally took used a different approach: Dedekind cuts (Richard Dedekind: 1831 - 1916). The two theories are equivalent, but it is a bit messy to prove it.

Cauchy Sequences

Let $x_1, x_2, x_3, \ldots$ be an infinite sequence of Rational Numbers, one for each Natural Number. One sequence is the constant sequence 3, 3, 3, ... three's forever. A second sequence is the Natural Numbers themselves: 1, 2, 3, ... Another is 1/2, 1/4, 1/8, 1/16, ... where the next element is half the current one. The only requirement here is that the numbers all be Rational. In some sequences, such as the previous one, the values get closer and closer to each other the farther you go out. We want to make that idea of "get closer and closer" mathematically precise.

Pick a small positive Rational Number. Traditionally small numbers have been named with lower case Greek letters. So call our number epsilon, ε. Suppose that once we go out far enough in our sequence, say past some index N, that $|x_k - x_n| < \varepsilon$ whenever $k>N$ and $n>N$. In other words, first decide how close you want elements to be and then find an N such that beyond that point *every* pair of elements is closer than the closeness measure you choose. Now we are getting close to what we mean. But, suppose further, that we can do this for *every* $\varepsilon > 0$.

In other words suppose that: For every Rational $\varepsilon > 0$ there is an $N \geq 1$, such that if $n>N$ and $k>N$, it is true that $| x_k - x_n | < \varepsilon$. In that case we will call the Rational sequence *Cauchy*. This absolute value is the distance between the elements named. Note that a constant sequence satisfies this, since the distance between elements is always 0, so if $\varepsilon > 0$, you can choose N = 1. Also note that, in general, the N you need depends on ε; the smaller the ε, the larger the N.

On the other hand, the sequence 1, 2, 3, 4, … is not Cauchy, since adjacent elements stay the same distance apart as you go out, so arbitrary pairs of elements can be arbitrarily far apart.

The sequence 1/2, 1/4, 1/8, 1/16, … is Cauchy since if N is a value such that $1/2^N < \varepsilon$, then beyond point N, any pair of elements differs by less than $1/2^N < \varepsilon$.

But you need to be careful. Suppose we merge a constant sequence (other than all zeros) with the above sequence, alternating members. For example, the sequence of all 1's. The result isn't Cauchy since no matter how far you go out, there are values farther out that are close to a distance 1 apart, so if $\varepsilon < 1$ you can't meet the Cauchy criterion.

Definition 9.1: If $\forall$ Rational $\varepsilon > 0$ $\exists$ $N \geq 1$ such that $\forall$ $n > N$, $k > N$, it is true that $| x_k - x_n | < \varepsilon$, then we call the sequence of Rational Numbers $x_1, x_2, x_3, \ldots$ a *Cauchy Sequence* of Rationals.

Mathematicians are likely to abbreviate it even more, with s.t. standing for "such that" and even a comma standing for "it is true that" but also for "and". We might express this definition as:

The sequence of Rational Numbers $\{x_n\}$ is called Cauchy provided that $\forall$ Rational $\varepsilon > 0 \; \exists \; N \geq 1$ s.t. $\forall \; n, k > N$, $| \; x_k - x_n | < \varepsilon$.

Note that the sequence $s = \{1/2^n \; | n \in \mathfrak{N} \}$ is both Cauchy and converges to zero as noted in the previous chapter. But those two ideas are not the same. A constant sequence 3, 3, 3,... is Cauchy but it doesn't converge to 0 (though we could define a broader notion in which it converges to 3). Here is another:

1, 1/2, 1/3, 1/4, 1/5, ... In general the *kth* element is $1/k$ for $k \geq 1$. This is also Cauchy and converges to zero.

Theorem 9.2: If $x = x_1, x_2, x_3, \ldots$ is a Cauchy Sequence, then it is bounded. That is, there is a Rational Number B>0 such that $| \; x_n \; | < B$ for every n.

Proof: First note that every finite set is bounded. Also note that $|a| - |b| \leq |a \pm b| \leq |a| + |b|$ for any Rational Numbers a and b. The symbol $\pm$ means you may either add or subtract the two values and the inequality will still be true.

Let x be a Cauchy Sequence. Choose $\varepsilon = 1$ and let N be such that $|x_n - x_m| < \varepsilon$ for every $n, m \geq N$. Then, in particular, $|x_n - x_N| < 1$ for every $n>N$. But then $|x_n| - |x_N| \leq |x_n - x_N| < 1$, so $|x_n| \leq 1 + |x_N|$ for every $n \geq N$. So the values beyond N are bounded, but there are only a finite number of elements prior to N. QED.

Definition and Properties

In order to define the real numbers, first let **R** be the set of Cauchy Rational sequences. It is a huge set, of course. Each element is a Cauchy sequence.

It isn't terribly difficult to see that the sum and difference of Cauchy sequences are themselves Cauchy. So **R** is closed under these operations. For example, if c and d are Cauchy sequences, choose an $\varepsilon>0$ and let N_1 be such that elements of c are within $\varepsilon/2$ beyond N_1. Let N_2 be such that elements of d are within $\varepsilon/2$ beyond N_2. Let N be the larger of N_1 and N_2. Then elements of the sum of c and d will be closer than ε beyond N. This follows from the fact that $|a + b| \leq |a| + |b|$.

Theorem 9.3: The sum and difference of Cauchy sequences are Cauchy.

Proof: Outlined above.

A bit harder is the following:

Theorem 9.4: The product of Cauchy sequences is Cauchy.

Proof: Let $x = \{ x_n \mid n \in \mathfrak{N}\}$ and $y = \{ y_n \mid n \in \mathfrak{N}\}$ be Cauchy Sequences. Since both x and y are bounded let B be a common bound for x and y.

The product is $xy = x_1y_1, x_2y_2, x_3y_3, \ldots$ Choose $\varepsilon > 0$ and let N be such that for all $n, m \geq N$, $|x_n - x_m| < \varepsilon/2B$ and $|y_n - y_m| < \varepsilon/2B$. Then, for $n, m \geq N$ we have $|x_ny_n - x_my_m| =$
$|x_ny_n - x_ny_m + x_ny_m - x_my_m| \leq |x_ny_n - x_ny_m| + |x_ny_m - x_my_m| =$
$|x_n(y_n - y_m)| + |y_m(x_n - x_m)| =$
$|x_n||(y_n - y_m)| + |y_m||(x_n - x_m)| <$
$B \cdot \varepsilon/2B + B \cdot \varepsilon/2B = \varepsilon$. QED

Theorem 9.5: If $\{x_n\}$ is a Cauchy Sequence then so is $\{|x_n|\}$.

Proof: This follows easily from the inequality $||a| - |b|| \leq |a - b|$.

We define a new relation in **R** and call it μ.

Definition 9.6. Let c and d be two elements of **R**. We say that $c \mu d$ provided that the sequence $c - d$ converges to zero.

Note that in this case d – c also converges to zero.

Theorem 9.7: The relation μ is an equivalence relation on **R**.

Proof: The reflexive and symmetric properties are *trivial* to show.

To show that μ is transitive, assume $x \mu y$ and $y \mu z$, where $x, y, z \in$ **R**.

Assume $x = \{ x_n \mid n \in \mathfrak{N}\}$, $y = \{ y_n \mid n \in \mathfrak{N}\}$, $z = \{ z_n \mid n \in \mathfrak{N}\}$.

Since $x - y$ and $y - z$ converge to zero, choose an $\varepsilon > 0$. Let N_1 be such that

$|x_n - y_n| < \varepsilon/2$ for all $n > N_1$. Let N_2 be such that

$|y_n - z_n| < \varepsilon/2$ for all $n > N_2$. Let N be the larger of N_1 and N_2.

Then $|x_n - z_n| = |x_n - y_n + y_n - z_n| \leq |x_n - y_n| + |y_n - z_n| < \varepsilon$ for all $n > N$. QED

The word *trivial* in mathematics means extremely easy. In the above, the result follows directly from the definition. But in general, saying something is trivial depends on your audience. There is a story from my days at the University of Kansas, that a professor, saying that a step in a proof he was presenting at the chalkboard was trivial, but being questioned on it by a graduate student, took about an hour of quiet reflection, and the examination of several books, to conclude that yes, it was, indeed, trivial. And so he announced.

Definition 9.8: The set $\mathfrak{R}$ of all of the equivalence classes of μ in R, is called the *Real Numbers*. We also say that an element of the equivalence class for $x \in \mathfrak{R}$ *converges* to x.

Thus, a Real Number is a set of Cauchy sequences, each pair of which has a difference that converges to zero. Also, note that the equality of Real Numbers is determined by the equivalence relation.

Note that earlier here we showed that $\sqrt{2}$ isn't Rational, but we also found a pair of Cauchy sequences whose successive elements were closer and closer to the $\sqrt{2}$ without bound. Those sequences are two of the (many) representatives in the equivalence class defining $\sqrt{2}$. The sequences we found depended on the starting point. We picked a value whose square was less than 2 and another whose square was greater than 2. The successive values *converged* to $\sqrt{2}$. Or, more precisely, their squares converged to 2. If you look at the squares of those numbers and take the difference between them and the constant sequence of all 2's (2, 2, ...) you will see that it converges to 0.

We asked earlier whether the integers "existed," and decided that they didn't outside of mathematics, since they form an infinite set. Individual integers can probably be said to exist in some sense. I have three sheep and you have five. Even Rationals have some real–world meaning: I have three fifths as many sheep as you do. But the Real Numbers are pure imagination. Each real number is represented by a huge number of infinite sequences. How huge? An infinite number of infinite sequences. If we were to express a real number that isn't Rational (such as) $\sqrt{2}$ in a decimal format it couldn't repeat itself at the right, no matter how far you look. Its decimal representation is, itself, infinitely long. We can only approximate it with a Rational, but can do so as closely as you like. This is pure mind–stuff.

Notation 9.9: When we write $x = E(\{x_n\})$ we indicate that $\{x_n\}$ is a Cauchy Sequence of Rational Numbers as well as the fact that x is the equivalence class containing that sequence.

Definition 9.10: Let x and y be Real Numbers and $x = E(\{x_n\})$ and $y = E(\{y_n\})$. Define the sum of x and y to be: $x + y = E(\{x_n + y_n\})$. Define the product of x and y to be: $x \cdot y = E(\{x_n \cdot y_n\})$.

Theorem 9.11: The sum and product of real numbers are well defined. That is, other sequences in the equivalence classes of x and y yield the same result.

Proof: Let $\{x_n\}$ µ $\{w_n\}$ and $\{y_n\}$ µ $\{z_n\}$. We must show that $\{x_n + y_n\}$ µ $\{w_n + z_n\}$ and $\{x_n \cdot y_n\}$ µ $\{w_n \cdot z_n\}$.

Proof for sum: Let $\varepsilon > 0$ be a Rational Number. Choose N such that for n > N it is simultaneously true that $|x_n - w_n| < \varepsilon/2$ and $|y_n - z_n| < \varepsilon/2$. Then $|(x_n + y_n) - (w_n + z_n)| = |x_n - w_n + y_n - z_n| \leq |x_n - w_n| + |y_n - z_n| < \varepsilon/2 + \varepsilon/2 = \varepsilon$.

Proof for product. Let B be a simultaneous bound for sequences $\{w_n\}$ and $\{y_n\}$. Let $\varepsilon > 0$ and N be such that for n > N, $|x_n - w_n| < \varepsilon/2B$ and $|y_n - z_n| < \varepsilon/2B$. Then $|(x_n \cdot y_n) - (w_n \cdot z_n)| = |(x_n \cdot y_n) - (w_n \cdot y_n) + (w_n \cdot y_n) - (w_n \cdot z_n)| = |(x_n - w_n) \cdot y_n + w_n \cdot (y_n - z_n)| \leq |(x_n - w_n)| \cdot |y_n| + |w_n| \cdot |(y_n - z_n)| < B \cdot \varepsilon/2B + B \cdot \varepsilon/2B = \varepsilon$. QED

Note on simultaneous limits and bounds. If each of two sequences satisfies some property involving a small $\varepsilon > 0$ for indices beyond some N. Then we can, by choosing the smaller ε value or the larger or smaller bound as appropriate and the larger N value, guarantee the property for both, or any finite number of sequences.

No matter how we try to define the Real Numbers, some things will be easy and others more difficult. The following theorem is key to understanding how the Cauchy sequence approach really does define the Real Numbers. Basically it says that if a Cauchy sequence has a tail that is bounded below by a definite positive Rational Number, then all equivalent sequences have tails that are also bounded below by some, possibly smaller, positive Rational Number. All the equivalent sequences might fall a bit below the original, but not by very much. Thus, we can conclude that if a Cauchy Sequences converges to a positive number then equivalent Cauchy Sequences have tails that don't get too close to zero for large n.

Theorem9.12: Let $\{x_n\}$ be a Cauchy Sequences. Suppose that ∃ a Rational Number $\mathcal{E}>0$ and a Natural Number N, such that $\forall\ n > N$, $\mathcal{E} < x_n$. Suppose $\{y_n\}$ is any Cauchy sequence equivalent to $\{x_n\}$. Then there is a Natural Number N_1, such that $\forall\ n > N_1$, $\mathcal{E}/2 < y_n$.

Proof: Let $\partial = \mathcal{E}/2$. Since $\{x_n\}$ and $\{y_n\}$ are equivalent, let N_0 be such that when n > N_0, $|x_n - y_n| < \partial$. So $-\partial < x_n - y_n < \partial$. With just the right hand inequality we have $\partial = \mathcal{E} - \partial < x_n - \partial < y_n$, for all n larger than N_1, the maximum of N and N_0. QED

Note that the implication is that if any Cauchy sequence is "bounded away from zero in the tail" by a Rational, then all equivalent Cauchy sequences are also "bounded away from zero in the tail" by a Rational that is not dependent on which sequence is chosen. This makes the following into a valid definition. And it is important to note that in the above proof, the inequalities are all between Rational Numbers. Many of the following proofs will have the same character as this one, though some are a bit more complex.

Definition 9.13: If $x \in \mathfrak{R}$ we say x > 0 provided that if x = $E(\{x_n\})$ then ∃ a Rational Number $\mathcal{E}>0$ and a Natural Number N, such that $\forall\ n > N$, $\mathcal{E} < x_n$. We say that x < 0 provided $-x > 0$.

We are trying to get at the fact that if two real numbers aren't equal, then the Cauchy Sequences that define them have tails that are actually separated by some positive value, which we can take to be a Rational Number. The sequences may "overlap" at first, but must eventually separate. But it takes a bit of work to make that precise. We start by comparing a Real Number to 0 and work from there. Note that every subsequence of a Cauchy Sequence is Cauchy since, in a subsequence (with elements renumbered), the position of an element of the subsequence is smaller than or equal to its position in the original.

Note that we are focusing especially on zero and that zero is really special here, since the definition of equivalence of sequences is in terms of convergence to zero. Likewise, our definition of $x < y$ can be rephrased in terms of a comparison with zero by using subtraction.

Theorem 9.14: A subsequence of a Cauchy Sequence is in the same equivalence class as the sequence containing it.

Proof: Let $\{x_n\}$ be Cauchy and let $\{y_n\}$ be a subsequence. Let $\mathcal{E}>0$ and choose N such that if n, m > N, $|x_n - x_m| < \mathcal{E}$. Since $\{y_n\}$ is a subsequence, the value y_n appears in the original sequence with an index no less than n. Therefore $|x_n - y_n| < \mathcal{E}$. Thus, the two sequences are equivalent. QED

Theorem 9.15: If x = E($\{x_n\}$) $\in \mathfrak{R}$ but $x \not> 0$ then $\forall$ Rational Number $\partial > 0$, $\{x_n\}$ has a subsequence $\{z_n\}$ such that $z_n < \partial$ for every n.

Proof: If $x \not> 0$, then $\forall\ \mathcal{E}>0$ and $\forall$ N, there is an n > N such that $x_n \le \mathcal{E}$. So choose $\partial > 0$ and let $\mathcal{E} = \partial/2$. Let N = 1 and choose n_1 > N such that $x_{n_1} \le \mathcal{E} < \partial$. Now, chose a new N > n_1 and then a new n_2 such that $x_{n_2} \le \mathcal{E} < \partial$. Proceed inductively to choose n_k, each of which has $x_{n_k} < \partial$. The x_{n_1} values form the desired subsequence. QED

Corollary 9.16: By the same reasoning If x = E($\{x_n\}$) $\in \mathfrak{R}$ but $x \not< 0$ then $\forall$ Rational Number $\partial > 0$, $\{x_n\}$ has a subsequence $\{z_n\}$ such that $z_n > -\partial$ (or, equivalently, $-z_n < \partial$) for every n.

Theorem 9.17: Let $\{x_n\}$ be a Cauchy Sequence and $\partial > 0$ be a Rational Number. Suppose that $\{x_n\}$ has a subsequence $\{ x_{n_k} \}$ such that $x_{n_k} > \partial > 0$ for every n_k. Then there is an $\mathcal{E} > 0$ and N such that $x_n > \mathcal{E}$ for every n > N.

Proof: Let $\mathcal{E} = \partial/2$. Then let N be such that for all n, k >N, $|x_n - x_m| < \mathcal{E}$. Since this is also true when x_m is in the subsequence, $-\mathcal{E} < x_n - x_{n_1} < \mathcal{E}$. Thus $x_n > -\mathcal{E} + x_{n_k} > -\partial/2 + \partial = \partial/2 = \mathcal{E}$ for n > N. QED

Theorem 9.18: If $x \in \mathfrak{R}$, then exactly one of the following is true: $x < 0$, $x = 0$, or $x > 0$.

Proof: Clearly, if $x = 0$ then neither $x > 0$, nor $x < 0$ can hold since x is defined by Cauchy Sequences that converge to 0. Likewise $x > 0$ and $x < 0$ are mutually exclusive. So, suppose that neither $x > 0$, nor $x < 0$. We need to show that $x = 0$ or, if x = E($\{x_n\}$) that $\{x_n\}$ converges to 0. Let $\mathcal{E} > 0$ and let $\partial = \mathcal{E}/2$. Choose a subsequence $\{y_k\}$ of $\{x_n\}$ such that $y_k < \partial$ for every k. Choose a subsequence $\{z_k\}$ of $\{x_n\}$ such that $z_k > -\partial$ for every k. Then, for every k, $y_k - z_k < 2\partial$.

Choose N_1 such that $|x_n - x_k| < \partial$ for every n, k > N_1. Now assume that k > N_1 and that x_k is an element of the subsequence $\{y_k\}$. Then $-\partial < x_n - x_k < \partial$, which implies that $x_n < \partial + x_k < 2\partial$ for every n > N_1. Using the sequence $\{z_k\}$ we may similarly show that $x_n > -2\partial$, so that $|x_n| < 2\partial = \mathcal{E}$ for large values of n. QED

Because of the Law of Trichotomy, Theorem 9.18, we can use the same statement definition of the absolute value for Real Numbers that we have seen before: $|x| = x$ if $x \geq 0$ and $-x$ otherwise.

Definition 9.19: If $x, y \in \mathfrak{R}$, we say that $x > y$ provided that $x - y > 0$. Similarly $x < y$ provided $y - x > 0$.

Corollary 9.20: If $x, y \in \mathfrak{R}$, then exactly one of the following holds: $x < y$, $x = y$, or $x > y$.

Proof: This follows from the above theorem by examining x – y.

Definition 9.21: We say $x \leq y$ provided $x = y$ or $x < y$.

Theorem 9.22: The relation $\leq$ is well defined in $\mathfrak{R}$.

Proof: Exercise. Use the fact that $\{x_n\}$ and $\{y_n\}$ are Cauchy to show that sequences equivalent to $\{x_n\}$ and $\{y_n\}$ yield the same result.

Note, however, that it is possible that for sequences $x = \{x_n\}$ and $y = \{y_n\}$, that $x_n < y_n$ for all large n but still have $x = y$. This was, in fact the case for the two sequences we used to "squeeze" $\sqrt{2}$. Likewise the "Zeno sequence" has all positive terms but converges to zero.

Theorem 9.23: The Rationals are embedded in a subset of $\mathfrak{R}$.

Proof: If r is a Rational Number, then it is represented by the Cauchy Sequence $\{r\} = r, r, r,...$

The sum and product of such sequences is clearly also rational. If $r \leq s$ then the sequence $\{r\}$ is less than or equal to the sequence $\{s\}$ as well, by the above definition. All the properties of sum, product and < apply to these sequences. QED

Definition 9.24: An element in $\mathfrak{R}$ that is not in $\mathfrak{Q}$ is called irrational.

We showed above that $\sqrt{2}$ is irrational. We will show below that most (by far) real numbers are irrational.

Note that one thing that we did not omit, is a definition of the quotient of Cauchy sequences in which the denominator converges to zero, but has no zero terms. That was intentional, and such quotients are extremely interesting, especially in Calculus and its extensions. And note that we can't claim that the quotient of Cauchy sequences is Cauchy. Indeed it might not be. For example, if you divide the constant sequence of all 1's by the sequence we examined for Zeno's Paradox you see that the resulting sequence has terms farther apart, not closer together. However, in order to define the quotient of reals, we do need to restrict the divisor to be a sequence that doesn't converge to zero.

Note the following. Suppose we have a Cauchy sequence, x, that converges to 0, so that it represents the Real Number 0. But suppose also that it has no zero elements. Then that sequence still has a multiplicative inverse, but that sequence can't be Cauchy so it doesn't represent any Real Number. To show that it isn't Cauchy, note that x has a monotone subsequence. Those elements are getting closer to zero, so their inverses (reciprocals) are getting farther apart.

Moreover, if a Cauchy sequence does not converge to zero then beyond some point it can have no zero elements. Therefore, by truncating if necessary we can find a sequence in the same equivalence class that has no zeros.

Theorem 9.25: Let $\{x_n\}$ be a Cauchy Sequence without zero elements that does not converge to 0. Then $\{1/x_n\}$ is also Cauchy.

Proof: Let $x = E(\{x_n\})$ with $x \neq 0$. Let $\varepsilon > 0$ be given. Let $\{y_n\}$ be the constant zero sequence. First, assume $x > 0$. Let $b > 0$ and N be such that $x_n > b$ for all $n > N$. Thus $|x_n| > b$ for all $n > N$. Thus $1/|x_n| < 1/b$ for all such n. Since $\{x_n\}$ is Cauchy let $N_1 > N$ be chosen such that for $n, m > N_1$ it is true that $|x_n - x_m| < \varepsilon/b^2$. Then $|1/x_n - 1/x_m| = |(x_n - x_m)/(x_n x_m)| = |x_n - x_m| / |x_n x_m| < \varepsilon$. A similar proof applies if $x < 0$. QED

Definition 9.26: Let $a = E(\{a_n\})$ and $b = E(\{b_n\})$ be Real Numbers with $b \neq 0$. Assume WLOG that $b_n \neq 0$ for every n. Define $a/b = E(\{a_n\} \cdot \{1/b_n\})$.

Theorem 9.27: The quotient defined above is well-defined.

Proof: It is enough to show that the sequence $\{a_n\} \cdot \{1/b_n\}$ is Cauchy. But this follows from the theorem above and the corresponding result for products.

We also need to show that addition and multiplication, as defined above are commutative and associative and that multiplication distributes over addition. But these are an easy consequence of the definitions. The sequences that define $x + y$ and $y + x$, for example, are the same sequences, term by term, because of properties of the Rational Numbers.

We extend the definition of Cauchy Sequences to sequences of Real Numbers in the obvious way. Up to now Cauchy sequences have been a sequence of Rationals.

Definition 9.28: A sequence $\{x_n\}$ of Real Numbers is Cauchy provided that $\forall$ real $\varepsilon > 0$ $\exists$ $N \geq 1$ such that $\forall$ $n, k > N$ it is true that $| x_k - x_n | < \varepsilon$.

Actually there is no need to permit ε to be real, as there is always a smaller Rational value that is also greater than 0. But now the question arises. If we repeat the construction using this new definition, do we get anything new? Or are the equivalence classes under the corresponding equivalence relation just exactly the real numbers? The answer to the last question is yes, so in this sense the reals are said to be *complete*. There are no additional "holes" in the number line.

Definition 9.29 We say that a sequence of Real Numbers $x = x_1, x_2, x_3, \ldots$ converges to a real number p, provided that $\forall \varepsilon > 0$ $\exists$ N such that $| x_n - p | < \varepsilon$ for every $n \geq$ N. We then write: $\lim_{n \to \infty} x_n = p$, which is read "the limit of x_n as n goes to infinity is p".

Theorem 9.30: (Completeness of the Real Numbers) A Cauchy Sequence of Real Numbers converges to a Real Number.

Here is a sketch of the proof. Suppose you have a Cauchy sequence x of Real Numbers. We will define two Cauchy sequences of Rational Numbers that are equivalent to each other, but squeeze x the way we squeezed a given number previously. Let $x = E(\{x_n\})$. Consider element x_n for a given n. This is a real number, so we may approximate it arbitrarily with Rationals. Let $y_n < x_n < z_n$ be such that $0 < z_n - y_n < 1/n$ with y_n and z_n Rational. The sequences $\{y_n\}$ and $\{z_n\}$ are Cauchy since their elements become arbitrarily close to elements of the Cauchy sequence x, and they are equivalent by construction. Therefore they converge to the same Real

Number, but then x, being squeezed between them, must also converge to this value. Because of this we say that the Real Numbers are *complete*. There are no additional holes to be filled by limits of real sequences.

Theorem 9.31: If A is a non–empty set of real numbers that is bounded above by a real number, then there is a real number b such that $x \le b$ for every x in A and for any real number t smaller than b, some element of A is larger than t. The value b is called the *least upper bound* of A. It may be in A or not.

Proof. Assume WLOG that the Rational Number U_0 is an upper bound of A. Since A is non–empty, there is a value y in A. Let L_0 be any Rational Number less than y. We will construct two Cauchy Sequences of Rational Numbers starting with these two values using the Squeeze Process that we saw above. Let m_0 be the average of L_0 and U_0. If m_0 is an upper bound of A let U_1 be m_0 and let L_1 be the same as L_0. But if m_0 is not an upper bound (some value in A is greater than m_0) then let L_1 be m_0 and U_1 be the same as U_0. Continue this process generating sequences L and U. Now every element of the U sequence is an upper bound of A, but none of the elements of the L sequence is. Since the two sequences are equivalent Cauchy sequences (i.e. their difference converges to 0 by construction) let their common value be b. It is an upper bound since every element in a sequence defining it is. It is the *least* upper bound since in the L sequence every term is less than some element of A. Thus, since the L sequence is increasing it will eventually be greater than or equal to any value less than b. QED

A corresponding theorem establishes *greatest lower bounds* for sets that are bounded below.

Example: The value 1 is the least upper bound of {x | x is a Real Number, x < 1}.

Theorem 9.32: Bounded monotone sequences of reals converge. Assume $x = x_1, x_2, x_3, \ldots$ is bounded and increasing, so that $x_n \le x_{n+1}$ for every n. (The argument for decreasing sequences is essentially the same.) Then considered as a set it has a least upper bound, p. The sequence then converges to p.

Proof: Let $\varepsilon > 0$. We need to show that $\exists$ N such that $|x_n - p| < \varepsilon$ for all n > N. In fact we need only show that for a single n, $|x_n - p| < \varepsilon$ since the sequence is increasing and p is an upper bound. Thus all further elements well be closer to

p. But suppose there is no such n. Then $p - x_n \geq \mathcal{E}$ for all n and so $p - \mathcal{E}$ is an upper bound of the sequence smaller than its least upper bound p. The contradiction establishes the result. QED

Lemma 9.33: If a sequence of Real Numbers $x = \{x_n\}$ converges to a real number $a \neq 0$, then $\exists$ N such that $x_n \neq 0$ for all $n > N$.

Proof: Exercise. Consider $\mathcal{E} < |a|/2$.

The real numbers have all of the following properties. We have proved most of them above. The others are not especially difficult, though most are a bit tedious. They depend on manipulating Cauchy sequences, of course.

(a) Addition is commutative and associative.

(*b*) Multiplication is commutative and associative.

(*c*) Multiplication distributes over addition.

(*d*) Zero is an additive identity: $0 + a = a$ for any real number a.

(e) One is a multiplicative identity: $1 \cdot a = a$ for any real number a.

(f) Every real number, a, has an additive inverse, $-a$. That is: $-a + a = 0$.

(g) Every non–zero real number, a, has a multiplicative inverse, $1/a$. That is: $(1/a) \cdot a = 1$.

(h) Substitution and Cancellation for addition. If a, b, and x are Real Numbers then $a = b$ if and only if $a + x = b + x$.

(i) Substitution law of multiplication: If a, b, and x are Real Numbers and $a = b$, then $a \cdot x = b \cdot x$.

(j) Cancellation law of multiplication. Let a, b, and x be Real Numbers with $x \neq 0$. If $a \cdot x = b \cdot x$ then $a = b$.

The less than or equal to relation, ≤, in $\mathfrak{R}$ satisfies:

(k) ≤ is reflexive

(l) ≤ is antisymmetric: if $x \leq y$ and $y \leq x$, then $x = y$.

(m) ≤ is transitive: if $x \leq y$ and $y \leq z$, then $x \leq z$.

(n) If $x, y \in \mathfrak{R}$ then $x \leq y$ or $y \leq x$. (The order is *total*).

(o) If $x, y, z \in \mathfrak{R}$ and $x \leq y$, then $x + z \leq y + z$.

(p) If $x, y \in \mathfrak{R}$ and x and y are both non–negative then $x \cdot y$ is also non–negative.

(q) $\mathfrak{R}$ has the least upper bound property given in Theorem 9.31.

Note that the Rational Numbers satisfy all of these, except the last.

(r) The real numbers are complete.

Many of them hold in the integers, as well. Some are true in the Natural Numbers. As we progressed through the extensions we added structure. Property (f) of course doesn't hold in the Natural Numbers and inspired our extension to the Integers. Similarly (g) led us to the Rational Numbers.

Cardinality of the Real Numbers

One of the ways to envision the Real Numbers is to use a line that extends infinitely in both directions. Somewhere on this line we mark a point for zero and somewhere, usually to the right, a mark for 1, establishing the "unit distance." Then we think of the integers marking off equally spaced segments on this line infinitely to the right as well as the left. The Rational Numbers are then filled in on the line proportionally according to their values, with 3/2 midway between 1 and 2 for example. Then the rest of the reals are imagined to fill the gaps according to the convergence of the Cauchy sequences defining them. Their distance from zero can be approximated with Rational Numbers that have powers of 10 as denominators, but the values of irrational numbers can't be given exactly this way. But we can use infinitely long decimal expansions to represent them, and approximate them with Rational Numbers to any desired degree of accuracy.

The set of Real Numbers is infinite, but $\mathfrak{R}$ is not countable. There are actually *more* Real Numbers than Rational Numbers in a precise sense. To see this, suppose that $\mathfrak{R}$ is countable. Then the reals strictly between 0 and 1 would be (at most) countable as well. But then we could list those in an infinite list, numbered by the Natural Numbers. Suppose we list the decimal expansions of the Real Numbers between 0 and 1 in some list where the farther we go down, the more decimals we write out explicitly. But each such Real is listed exactly once. Note that each number is 0 to the left of the decimal point and extends infinitely (in theory) to the right. We are going to find a Real Number not in this list, contradicting the statement that we had listed all of them. The digits of the constructed number will be only 7's and 8's. Note that we also need to replace any (Rational) number in this list that ends in all 9's with an equivalent representation that ends in all 0's, similar to what we did earlier.

To build this new number look at the first digit to the right of the decimal point of the first number in the list. If it is an 8, write down 7 as the first digit of our new number, but otherwise write 8. Now look at the second digit of the second value in the list. If it is an 8 write 7 for the next digit, otherwise write 8. Continue this process, writing a number consisting of just 7's and 8's. Note that the number you are writing is not equal to any of the numbers in the list since it differs from the n'th value in the n'th decimal place. It is, in fact "quite a lot different" from all of the values above it in the table. But this is contrary to our assumption that the list contained all of the reals.

Therefore, the cardinality of $\mathfrak{R}$ is greater than that of $\mathfrak{N}$. The cardinality of $\mathfrak{R}$ is normally written as $\mathfrak{c}$. There is actually a theory of transfinite arithmetic that is interesting to study. It is at the edges of Set Theory in that you need additional Axioms beyond those normally used in set theory to prove some of the theorems. For example, we have two infinities defined now, that of the Natural Numbers and that of the Real Numbers. So the question arises, are there any in between? Georg Cantor studied these questions extensively, but not exhaustively, in the 1880's.

Another extension to the concept of an infinite series, this time in Set Theory, is to interpret the + sign in a series as set union. Then a union of an infinite sequence of sets can be studied. For example, define a set R_x to be the set of Rational Numbers less than the Rational Number x. Then consider the set of such sets for which $x < 0$ or $x^2 \leq 2$. Then we can alternately define the number $\sqrt{2}$ as the union of all these sets. This is similar to the idea of a Dedekind Cut mentioned earlier. We won't explore that idea here, however. But most ideas in maths can be extended and the extensions explored.

Foreshadowing: Topology

Some of the properties of the Real Numbers have been generalized to create a new field of study: Topology. In $\mathfrak{R}$, an ε–*neighborhood* of a value x, for $\varepsilon>0$, is defined by

$N_\varepsilon(x) = \{ y \mid y \in \mathfrak{R} \text{ and } |x - y| < \varepsilon\}$.

In other words, the set of real numbers *close* to x as measured by ε.

An *open set* in $\mathfrak{R}$ is a set, A, that itself contains an ε–neighborhood of each of its points, x, for some ε which may depend on x. In other words, A is *open* provided that $\forall\, x \in A\ \exists\ \varepsilon > 0$, such that if $|y - x| < \varepsilon$, then $y \in A$.

In particular, given Real Numbers a, and b, the set $\{x \mid a < x < b\}$ is open. For a given x, the smaller of the distance between x and a and between x and b is a suitable ε. Such a set is called an *open interval*. This open interval is written (a, b), but the meaning is not the same as the simple ordered pair.

It isn't difficult to show that an arbitrary union of open sets is open, nor that a finite intersection of open sets is open. The complement of an open set is called *closed*. The set of Real Numbers itself is open and by convention the empty set is also. It is easy to show that for given a and b, that $\{x \mid a \leq x \leq b\}$ is closed. It is called a *closed interval*. The closed interval is written $[a, b]$.

Note that the set of Rational Numbers, as a subset of $\mathfrak{R}$, is neither open nor closed. And $\mathfrak{R}$ itself is a both open and closed. The closed interval from a to b isn't open since it contains no neighborhood of either a or b, though it does contain neighborhoods of its other (interior) points. Nor is the open interval from a to b also closed, since its complement isn't open, not containing neighborhoods of a or b.

These ideas have been generalized as follows. Consider a set X. A *topology* on X is a collection, τ, of subsets of X with the following properties (axioms).

(a) $\varnothing \in \tau$

(b) $X \in \tau$

(c) The union of any arbitrary collection of elements of τ is in τ

(d) The intersection of any finite collection of elements of τ is in τ.

The elements of τ are called *open sets*. Their complements in X are called *closed*. A set with a topology is said to be a *topological space*. So $\mathfrak{R}$ is a topological space using the above definitions of open sets there. The elements of the underlying set, X, of a topological space are normally called *points*, since many of the interpretations of topologies are geometric in nature. One simple, and trivial, example of a topology on an arbitrary set has every subset open. This is called the *discrete* topology.

Topology is the study of topological spaces. Note that it focuses on sets of points, unlike Abstract Algebra, which focuses on operations.

Since the time I was in college, an important and long-standing theorem in topology was proven. We mentioned this before. The four-color theorem(+) says that it is possible to color maps in the plane (or on a sphere), so that adjacent regions have different colors, using only four colors. It was a problem nearly every mathematician my age once played with. The first proof was provided partly with the use of a computer as it has a huge number of cases.

Foreshadowing – Calculus and Analysis

One definition of Mathematical Analysis is that it is the study of *continuous* functions (discussed below). Calculus is a subfield of Analysis. In calculus you study the ideas of the *differential* and the *integral*, but in fact, these are used to study the relationship between what happens locally to a function (behavior near a given real number) to the global or overall behavior. One view of a real valued function of a real variable (i.e. both domain and range are in $\mathfrak{R}$) is a set of ordered pairs. If the function is named p then the pairs are {(x, p(x) | x is in the domain of p}. We can graph pairs, of course. Here is the graph of some of the points from the function $f(x) = x^2 - 2x + 2$. A few of these points are (0, 2), (1, 1), (3, 5), etc. This is a continuous function everywhere, as are all polynomials.

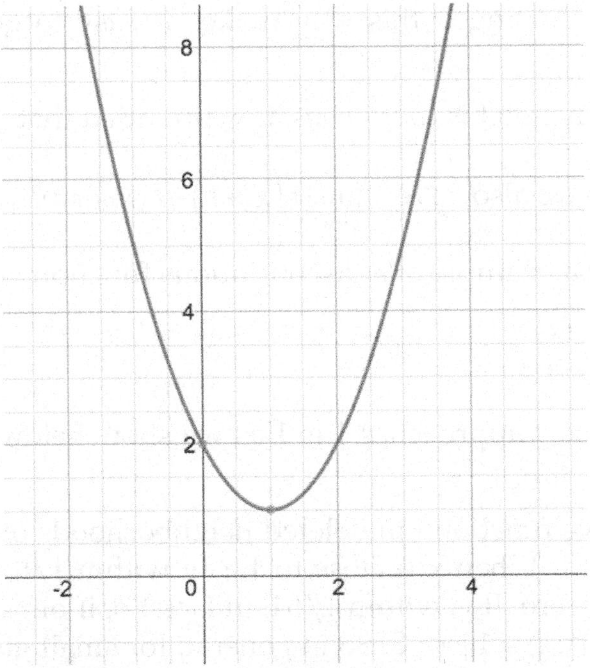

Definition 9.34: A function f from $\Re$ to $\Re$ is said to have limit L as x approaches x_0, provided that $\forall\, \varepsilon > 0\ \exists\, \partial > 0$ such that $|f(x) - L| < \varepsilon$ whenever $0 < |x - x_0| < \partial$. We then write $\lim_{x \to x_0} f(x) = L$, which reads "the limit of f(x) as x approaches x_0 is L".

Since x_0 is fixed we also write, by substituting $x_0 + h$ for x, $\lim_{h \to 0} f(x_0 + h) = L$. The implication, is that if x is near x_0 then so is $x_0 + h$ for small values of h.

Previously we have only looked at limits of *sequences* as the index approaches infinity. Now, however we can look at the limit of a general function as real numbers approach a given value. Notice that the inequality $0 < |x - x_0|$ means that $x = x_0$ is excluded from consideration. The inequality $|x - x_0| < \partial$ means that x is close to x_0. We could restate this in terms of neighborhoods. Call the set $\{x \mid 0 < |x - x_0| < \partial\}$ a *deleted neighborhood* of x_0. Then the $\lim_{x \to x_0} f(x) = L$ provided that for any neighborhood N of L there is a deleted neighborhood N_0 of x_0 such that f(x) is in N for all x in N_0. Stating it this way makes it easy to generalize the idea to Topological Spaces.

Definition 9.35: A function f is *continuous* at x_0 provided that $\lim_{x \to x_0} f(x) = f(x_0)$.

Note that in this case we also have $\lim_{h \to 0} (f(x_0+h) - f(x_0)) = 0.$

A somewhat contrived example of a discontinuous function is

$$f(x) = \begin{Bmatrix} x^2 - 2x + 2, if\ x \neq 1 \\ 0, \qquad if\ x = 1 \end{Bmatrix}$$

This function's limit as x approaches 1 is 1 as we show below. But the value there is 0 as set explicitly.

If we look at the above function f in deleted neighborhoods of x = 1, it has formula $x^2 - 2x + 2 = (x-1)^2 + 1$. When x is close to 1, say within 1/8, the first term of this (i.e. $(x-1)^2$) is close to zero. It is within 1/64, in fact. So, if our chosen ε is 1/64, we can let ∂ be 1/8 and $|f(x) - 1| < \varepsilon$. So in general, for small given ε the ∂ can be as large as the square root of ε.

An even simpler function, that has no limit at all at $x_0 = 0$ is:

$$g(x) = \begin{Bmatrix} 0, & if\ x \leq 0 \\ 1, & if\ x > 0 \end{Bmatrix} .$$

As you approach 0 from the left side the value stays 0. As you approach from the right it is just 1. So in any deleted neighborhood of 0, the function will have values that differ by exactly 1. So no ε neighborhood of any value L will guarantee the function is close to L if $\varepsilon < 1$. Having no limit, the function cannot be continuous at 0. The function g(x) is an example of a *step function*.

We will leave the exploration of continuous and discontinuous functions to your Calculus course.

Polynomials, such as the first example above are very well behaved functions. But poorly behaved functions are more interesting to study. If you take the function $f(x) = x^2 - 2x + 2$ and modify the left part of it by vertical reflection, you get the function:

$\begin{cases} x^2-2x+2, if\ x>1 \\ -x^2+2x,\ if\ x\le 1 \end{cases}$ The graph is shown below. The graph looks perfectly ordinary, just as smooth as the original polynomial. But it isn't. It has a strange property at x = 1 that you can explore when you know about differential Calculus. Informally, however, note that if you were driving along a road shaped like this curve, from left to right, you would need to turn harder and harder to the right as you approach x = 1, but then, without any transition at all, you would need to immediately turn hard to the left.

Here is a quick look at the derivative, one of the other key parts of mathematical analysis. Suppose we pick a point on a curve. For the original function above, let's look at the point (2, 2). If $f(x) = x^2 - 2x + 2$, then f(2) = 2, so this point is on the curve. We want to ask what direction the curve is "moving in" at this point. In other words, if a car were driving along a road shaped like this, what direction would the car be going at that point. The meaning of the question needs to be examined carefully, since the car is actually turning. However, if the road is straight, a line, then the answer is easier. The slope of the line, which is the ratio between the amount of vertical change to the amount of horizontal change, is a pretty good definition. The change in a variable is often called a delta, so this slope is delta y divided by delta x at that point. That is: δy/δx. For a straight line, this is the same everywhere so it is easy to compute. Suppose the line contains two points, say the point (2, 2) and the point (5, -3). Between these two points the change in y is δy = -3 – 2 = -5 and the change in x is δx = 5 – 2 = 3, so the slope is -5/3. This is the direction of the line at that point and at every point. An equation of the line is then y = (-5/3)x + (16/3). This comes from the "point-slope" form for the equation of a line. A line containing (5, -3) with slope -5/3 is $y-(-3)=\frac{-5}{3}(x-5)$.

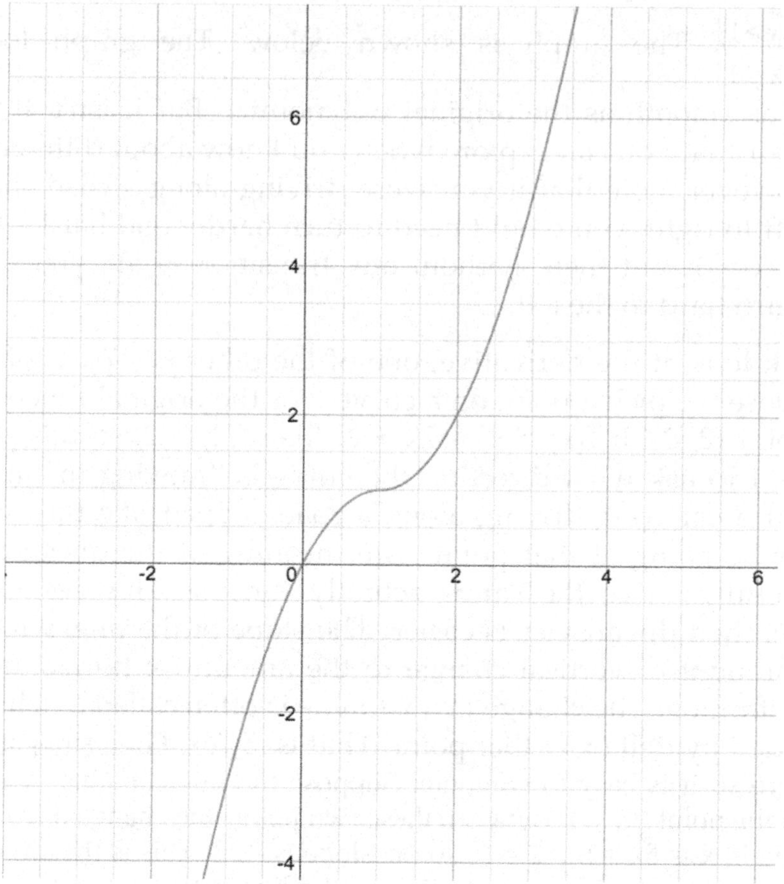

For a curve, it isn't quite so simple, of course, so instead, we take the direction of the curve to be the direction of the tangent line to the curve at that point. But not every curve has a tangent line so this is where the derivative comes in. The tangent will exist when the derivative does and the tangent line itself is defined to be the line passing through the point whose slope is the derivative. So it is quite important. Essentially, the derivative is a limit of a certain quotient. In math that is more advanced than what is taught in early calculus, the idea of limit might be as defined above or more generalized. Likewise the quotient might be just as we see below or something more complex. But the idea remains. A “limit” of a “difference quotient”.

Definition 9.36: Suppose the function f(x) is defined in an open interval containing x_0. The derivative of f at x_0 is defined to be $f'(x_0) = \lim_{h \to 0} \frac{f(x_0+h)-f(x_0)}{h}$ provided the limit exists as a finite number.

Note that since the denominator has a zero limit, so must the numerator. This implies that if the derivative exists, then f must be continuous at x_0. Also note that h may be positive or negative here and that the numerator is the vertical change in the value of the function over the horizontal change of size h at x_0. The difference quotient inside the limit here is, therefore, the slope of a line that meets the graph of f at x_0.

For example, let $f(x) = x^2 - 2x + 2$, as above, and let x_0 be 2. Note that f(2) = 2 as you can see by evaluation or examination of the graph above. If we draw a line through this point, intersecting the graph it might look like the following:

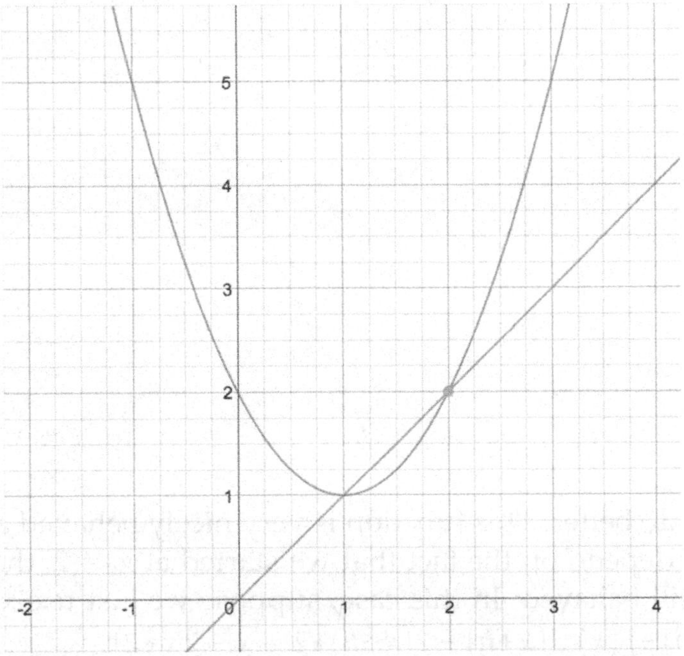

This line isn't tangent to the curve and is called a *chord*. The slope of the line is 1. But look at it this way instead. Let h = -1. Then the numerator of the difference quotient is f(2 – 1) – f(2) = 1 – 2 = -1 and the denominator is also -1, giving the slope. So, the difference quotient is the slope of a chord of the curve with one point fixed at x_0.

Now imagine what happens as h gets closer to zero. The chord line will get steeper in this example. In fact, the difference quotient for this function at $x_0 = 2$ is $\frac{f(2+h)-f(2)}{h}=\frac{(2+h)^2-2(2+h)+2-2}{h}=\frac{h^2+2h}{h}=h+2$. When h = -1 this is 1 as we just saw. But it is also true that when h is close to zero, this slope is close to 2. Therefore the derivative is 2 at $x_0 = 2$ and we can see the tangent line (with slope 2) in the following:

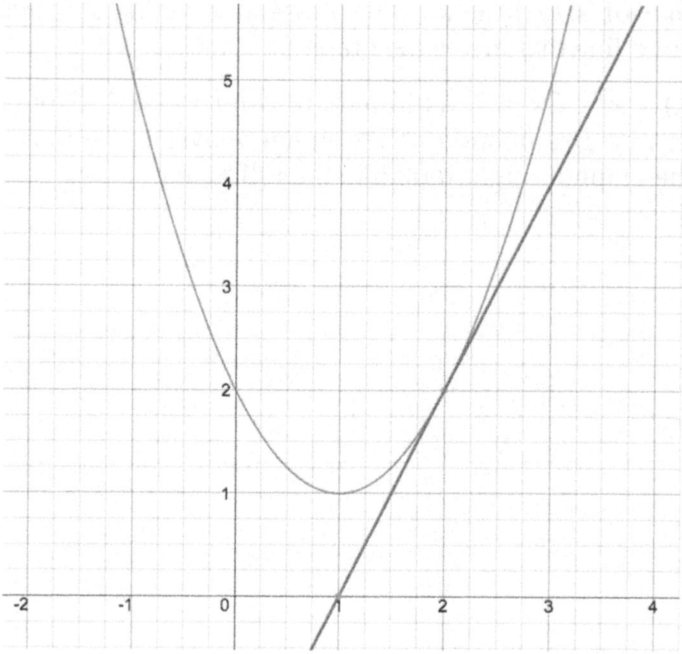

However, we can do better. This function is very nicely behaved and the essence of the above doesn't depend on the fact that we started at $x_0 = 2$, though it would for a function that is ill behaved. In this case, suppose we just use $x_0 = x$ instead. We then have: $\frac{f(x+h)-f(x)}{h}=\frac{(x+h)^2-2(x+h)+2-(x^2-2x+2)}{h}=h+2x-2.$

The limit of this as h approaches zero is $f'(x) = 2x - 2$ for any x, so we can find the derivative of the original function at any point x. Again note that this is 2 when x = 2 as we saw above. But you can look at this and find tangent lines anywhere. For example, when x = 1 this is zero, so the tangent line is horizontal. This is where the function has a minimum, an important observation. To summarize, the derivative of a function at a point tells us the direction of the curve at that point, even though it is changing as you move along. Also note that when the derivative is defined in some region then it makes sense to talk about the derivative of the derivative, or the second derivative, etc. Each such derivative tells you something different about the original function. The second derivative is related to the curvature of the original: how fast it is turning and in which direction. The derivative is a power tool for examining the behavior of functions.

I have greatly enjoyed studying the behavior of misbehaved functions. Here is one that is very misbehaved in neighborhoods of 0. In fact it assumes every real value in every neighborhood of 0. It is functions like this that make the study of mathematics interesting.

Let $f(x) = (1/x)\sin(1/x)$. Part of the graph follows. Note that the figure is truncated on all sides as it extends upward and downward more and more as you approach $x = 0$. It oscillates up and down so much that it seems to fill in the space. But even this function is very well behaved as long as we stay away from 0. That is, if you work in neighborhoods that exclude 0. Again, differential calculus is the tool of choice.

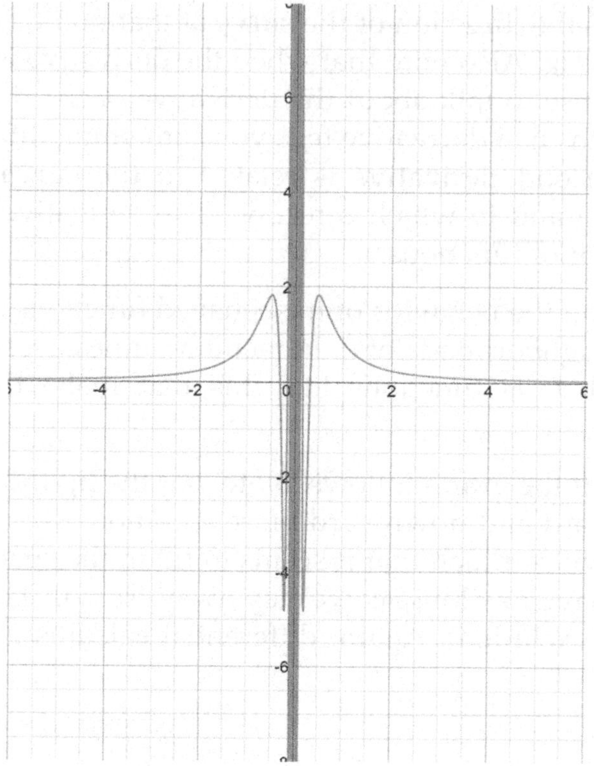

This function has the interesting property that if y is any Real Number at all then the function will have value y somewhere in every neighborhood of x = 0. In fact, it will take the value y infinitely often in any neighborhood of x = 0. Another way to say it is that if you draw a horizontal line anywhere in the above graph, the function will cross that line infinitely often.

Another misbehaved function is one that is equal to zero for any irrational number and equal to one for any rational. This function is continuous nowhere, so is very misbehaved. But there are lots of possibilities. For example, the following function is continuous at x = 0, but nowhere else.

$$f(x)=\left\{ \begin{array}{ll} x & \text{if } x \text{ is rational} \\ 0 & \text{if } x \text{ is irrational} \end{array} \right\}.$$

Summary So Far

We have seen that, in a way, the Real Numbers complete the Rational Numbers by filling in the holes that the Rationals leave in the number line. We showed one development, using Cauchy Sequences, and it may seem a bit complicated. The other approaches are also complicated in various ways. But then, the Real Numbers themselves are a deeply abstract concept, as should be obvious by now. That, by the way, is a good thing to a mathematician. Abstraction is good. It is our big hammer, after all.

Since we didn't need to discuss it for the development above we haven't mentioned that there are actually two square roots of two, one positive and one negative. We started developing our Rational estimating sequences with numbers close to the positive square root, but could have done so starting near the negative value just as well.

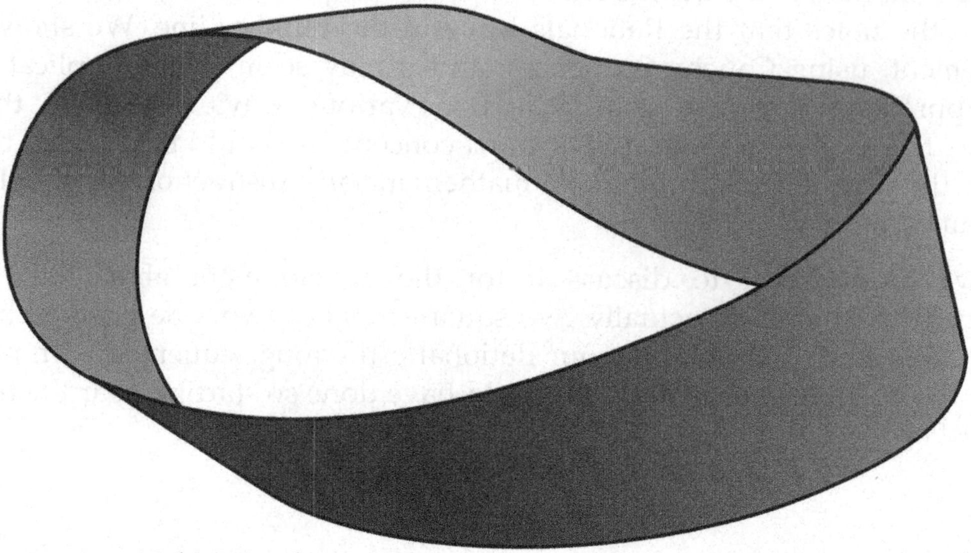

Möbius Band

10 Polynomials

Definition and Properties

We will describe polynomials of a single variable here, though they have extensions to multivariate polynomials as well. If x represents a real number then we can define x^n for $n \geq 0$ inductively. Define $x^0 = 1$ and $x^{k+1} = x \cdot x^k$. The value n in x^n is called the exponent. We can also define x^n for negative exponents as: $x^{-1} = 1/x$ provided $x \neq 0$ and $x^{-(k+1)}$ as $x^{-(k)}/x$. We won't need the negative exponents here.

Definition 10.1: A *real polynomial of degree n*, where n is a Whole Number, is a function (here with both domain and range in $\mathfrak{R}$) of the form

$f(x) = a_n x^n + a_{n-1} x^{n-1} + \ldots + a_1 x + a_0$ where the a's are real constants, called *coefficients*. We assume $a_n \neq 0$ unless n = 0.

A polynomial of degree 0 is called a *constant polynomial*. For a constant polynomial only a_0 can be non-zero. It is just a function that always returns the same real number. If all of the coefficients are zero except for one of them, the polynomial is called a *monomial*. For example $3x^5$ is a monomial.

Definition 10.2: A *root* of a polynomial is a value of the independent variable that yields 0 for the value of the function. In other words, a root of f is a solution to $f(x) = 0$. A constant polynomial has no roots at all unless it is the constant 0, in which case every Real Number is a root. This case is considered degenerate and not otherwise considered.

Polynomials have long been important in math and we have used several of them above. For example $g(x) = x^2 - 2$ was used to show why we have holes in the Rational Numbers. This polynomial has no Rational root. We used other polynomial functions to motivate the Integers and Rationals as well.

A polynomial of degree n has potentially n roots. For example $f(x) = x^2 - 2x - 3$ has roots 3 and –1. Evaluate the function for either of these and get 0. We can factor this polynomial into two polynomials of degree 1 using the roots. $f(x) = (x - 3) \cdot (x$

+ 1). The polynomial $f(x) = x^2 - 2x + 1$ has only one real root, but the factorization is $(x-1)\cdot(x-1)$ and so we say that the root, 1, is repeated, giving us 2 roots, counting multiplicities.

Lemma 10.3: Every real number is a root of some polynomial with real coefficients.

Proof: Trivial.

Lemma 10.4: Any root of a polynomial with Rational coefficients is also the root of a polynomial with integer coefficients.

Proof: Trivial.

We have already seen that some irrational Real Numbers are roots of polynomials with integer coefficients.

Definition: A Real Number is called *algebraic* provided it is the root of some polynomial with integer coefficients. Otherwise it is called *transcendental*.

So $\sqrt{2}$ is algebraic. So is every Rational Number. It took a long time to discover that not all reals are algebraic. In fact, *most* of the real numbers are transcendental.

However, the polynomial $g(x) = x^2 + 1$ has no real roots at all.

We would like to have a system in which all polynomials of degree n have n roots, counting repeated roots. But we aren't there yet. Nor can we tell the complete story of algebraic numbers as the above polynomial also indicates.

But even among the real numbers, some are known to be transcendental. For example the number pi, or π, is not algebraic. While most reals are transcendental, few are actually known, as it is difficult to prove whether a given number is algebraic or transcendental. The number π is the ratio between the circumference and the diameter of a circle, but can also be defined by an infinite series

$$\pi = 4\cdot\sum\{(-1)^{n+1}/(2n-1) \mid n \in \mathfrak{N}\} = 4\cdot(1 - 1/3 + 1/5 - 1/7 + \ldots).$$

Likewise Euler's Number e, the base of the natural logarithms, is transcendental. It's value is (note that 0! is 1, as noted before):

$$e = \sum\{1/n! \mid n \in \mathfrak{W}\} = 1 + 1/1 + 1/2 + 1/6 + 1/24 + \ldots.$$

It is possible to show that the algebraic numbers are countable. Recall that a number is algebraic if it is the root of some polynomial with integer coefficients. The proof is to show that the number of such polynomials is countable and each has a finite number of roots. Therefore, the transcendental numbers are not countable. One can also show that the transcendental numbers are dense in the Real Line as well. One would expect that, since there are so many of them, but it isn't obvious. The result is a bit odd, however, since it is hard to even show whether any given number is algebraic or transcendental.

Let P be the set of Polynomials with real coefficients. Addition and multiplication are easily defined using the usual rules of operations on real numbers (commutative, distributive, etc). The constant polynomial with coefficient 0 is an additive identity in this set and the constant polynomial with coefficient 1 is the multiplicative identity. The additive inverse of a polynomial is the one with the same degree, but with all of the coefficients reversed in sign. With these operations, P is a Commutative Ring. However, almost none of the members of P have multiplicative inverses. If the coefficients are drawn from fields other than the Real Numbers, a polynomial might well have an inverse. Mathematics is wonderful.

Torus

11 The Complex Numbers

In Chapter 9 we noted at the end that as we have progressed here we have added properties as we went along. But we can also take a system and explore what might happen if we drop some property. This is another useful technique for math. Suppose we don't require a totally ordered number system. Hmmm.

In the real numbers we can solve many more equations than we could solve in the Rationals. The Rationals have no solution for $x^2 = 2$, for example. In the reals we can solve that, but also things like $x^2 - 2x - 3 = 0$. They aren't always easy to solve, but the solutions are easy to check. Here –1 or 3 will do. But we can't solve all such equations. One of the simplest is just $x^2 = -1$. But suppose...

Definition and Properties

Suppose we imagine a system that contains the real numbers, but in which $x^2 + 1 = 0$ has a solution.

Definition 11.1: Let $\mathfrak{C} = \mathfrak{R}^2$, the set of ordered pairs of real numbers. Define the following operations and relation:

If $r = (w, x)$ and $s = (y, z)$ are in $\mathfrak{C}$, then

$r + s = (w + y, x + z)$

$r \cdot s = (wy - xz, wz + xy)$

r = s iff $w = y$ and $x = z$.

We embed $\mathfrak{R}$ into this set by letting a real value x map to $(x, 0)$. Note that $(0, 1) \cdot (0, 1) = (-1, 0) = -1$ using the definition of multiplication just given. So in $\mathfrak{C}$, $x = (0, 1)$ is a solution to $x^2 = -1$. The reader should check that the properties of addition and multiplication of Real Numbers is preserved in the subset corresponding to the reals using the definitions above.

The set $\mathfrak{C}$ is called the **Complex Numbers**. A more common notation for (x, y) in $\mathfrak{C}$, however is x + iy, where i is a new symbol and + isn't yet associated with addition as defined above. The i comes from *imaginary*, and numbers of the form 0 + ix are called pure imaginary. The name is historical and, as you can guess, there was some controversy when they were first proposed. "Good grief, man. *Those* numbers aren't real. They are *imaginary*." Well, yes. Yes they are.

The first coordinate of (x, y) is called the real part and the second is called the imaginary part. The terminology carries over to the x + iy representation.

But the ordered pair representation is also useful. One way to think of a complex number is as a point in a coordinate plane. The horizontal axis is the "real axis" and the vertical is the "imaginary axis". A complex number can also be thought of as a vector from the origin to that point.

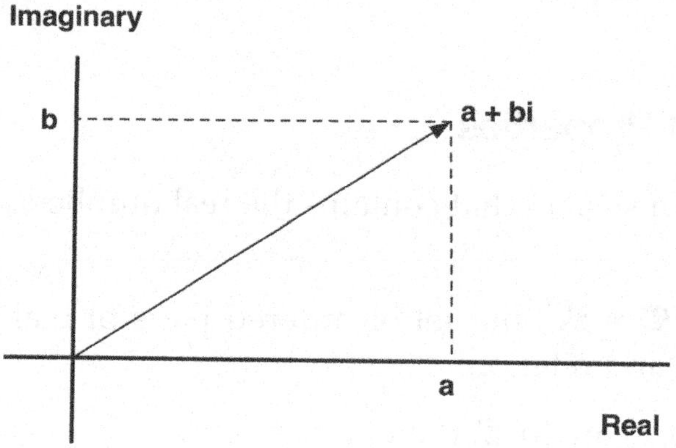

But if we let i = (0, 1), then, if x and y are real, we see that x + iy = $(x, 0) + (0, 1) \cdot (y, 0)$. = $(x, 0) + (0, y) = (x, y)$. Here we did use the addition of $\mathfrak{C}$ as defined above, so the "+" in x + iy does, indeed mean addition. And, of course, i = $\sqrt{-1}$, since it solves $x^2 + 1 = 0$.

Just as the Real Numbers can be visualized as an infinite line, the Complex Numbers form a plane. The positive real parts are to the right and the positive imaginary parts are at the top.

$\mathfrak{C}$ obeys some but not all of the properties of the reals. In particular, there is (and can be) no natural ordering in the complex numbers that is consistent with addition and multiplication, so none of the rules of inequalities carry over, including the least upper bound property. We can, and do, however, define < for the subset that has 0 as the second component: $(x, 0) < (y, 0)$ iff $x < y$. We need this in order to complete the embedding of $\mathfrak{R}$ into $\mathfrak{C}$.

Note, however, that the subset in which the first component is zero, the pure imaginary numbers, don't behave like a copy of the reals. In particular, the product of two imaginary numbers isn't imaginary, but real.

On the other hand, $\mathfrak{C}$ has one very important property that the reals do not. Polynomials in $\mathfrak{C}$ with integer, or even complex, coefficients have roots. So if we are just interested in solving equations involving polynomials, they have solutions, so there is no need to extend further.

Theorem 11.2: Fundamental Theorem of Algebra: A complex valued polynomial of degree n with complex coefficients has n complex roots, counting multiplicities. For this reason the Complex Numbers are said to be *Algebraically Closed.*

We won't prove that here. You will need to wait for a course in complex analysis[(+)]. Nor will we prove the other results we state here, saving that for your future study.

The algebra of Complex Numbers can be carried out with just the normal rules we have seen before, such as the associative, commutative, and distributive laws, along with the fact that $i^2 = -1$. The last implies that $i^3 = -i$ and $i^4 = 1$, of course. However a few shortcuts will help. The *complex conjugate* of $z = a + bi$ is the number $\bar{z} = a - bi$. In the figure above it is represented by the point that is the reflection of z in the real axis. But $z \cdot \bar{z} = a^2 + b^2$, a Real Number. Moreover the square root of this number is called the *magnitude* of z, which is written $|z|$. So $|a + bi| = \sqrt{a^2+b^2}$. Note that this is the distance from the origin to the point a + bi in the diagram above, by the Pythagorean Law.

You need to be a bit careful, however, since $|i| = 1$, it turns out that $|i|^2 = 1$ but $i^2 = -1$. That is not what you expect from studying real numbers.

Likewise, we can define the *distance* between two complex numbers $x = a + bi$ and $y = c + di$ as $|x - y| = \sqrt{(a-c)^2+(b-d)^2}$.

The *reciprocal* of a non-zero Complex Number $z = a + bi$ is $1/z = \bar{z} / (z \cdot \bar{z})$, or $(a - bi)/(a^2 + b^2)$, which has real part $a/(a^2 + b^2)$ and imaginary part $- bi/(a^2 + b^2)$. The quotient of two Complex Numbers can then be defined in terms of multiplication by the reciprocal.

Since we have a (real valued) distance function between complex numbers, we can define neighborhoods. These are like the open intervals in the Real Numbers, but are circles in the complex plane.

Definition 11.3: An ε-neighborhood, $N_\varepsilon(p)$ of a complex number p is $\{ z \in \mathfrak{C} \mid |z - p| < \varepsilon \}$. A deleted neighborhood is $\{ z \in \mathfrak{C} \mid 0 < |z - p| < \varepsilon \}$.

Given neighborhoods in $\mathfrak{C}$ we can define a Topology of open sets and also the continuity of functions, along with derivatives.

An open set in $\mathfrak{C}$ is one that contains a neighborhood of each of its points. Note that an open interval in $\mathfrak{R}$, isn't open in $\mathfrak{C}$ since it contains no points off the real line. In fact the real line (the set $\mathfrak{R}$) is closed in $\mathfrak{C}$ but not open. The study of the properties of complex valued functions is extremely rich, but we won't explore it here. It has certain features that differ substantially from those of real valued functions.

Foreshadowing – Trigonometry

I suppose that it is likely that you have already studied Trigonometry by now. I'll show a few things of interest to show how "trig" is connected to the Complex Numbers. Trig was developed in ancient Greece as the study of triangles, especially the angles and lengths of the sides. The sine and cosine functions, along with the others, have many modern applications in, for example, electrical engineering. The figure above, showing the geometric interpretation of a Complex Number is a good jumping off point for our discussion. Here is the same figure, but with some additional information.

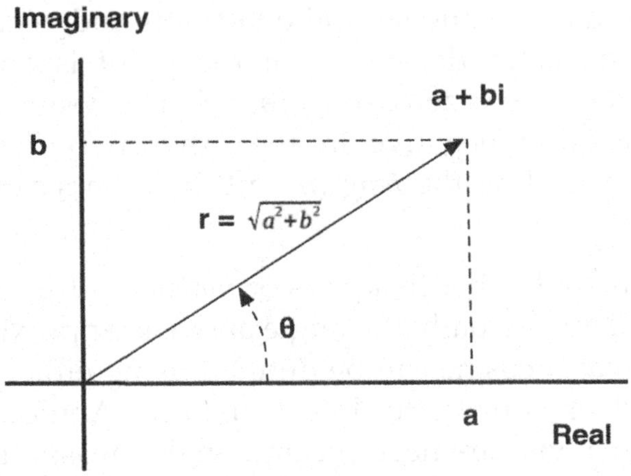

The same Complex Number, z = a + bi is shown, again as a vector, but this time we also show the (minimum positive) angle, θ, between the real axis and the vector. The vector and angle determines a right triangle of which the vector is called the hypotenuse. One side is along the real axis and the other is parallel to the imaginary axis. In the figure, we also indicate the length, r, of the vector, which is actually just $|a + bi|$.

The sample point a + bi is in the upper-right quadrant of the complex plane, called the first quadrant. Quadrants are numbered counterclockwise from there, so the one below the first is the fourth. The vertical dashed line in the figure, of length b, is called the *opposite* side of the triangle since it is opposite the angle. The segment from the origin to the point marked a is called the *adjacent* side. The sine of the angle, sin(θ), is normally defined to be the *length* of the opposite side divided by the length of the hypotenuse, or b/r here. However, note that a and b have signs. In the first quadrant both are positive. If the point were in the fourth quadrant then b would be negative, and the sine would have a negative value. Thus, our definition of sin(θ) as b/r, where b is the imaginary component of the Complex Number, is more explicit, taking signs into account. The hypotenuse, r, is always taken to be positive, however.

The cosine, cos(θ) is the *signed length* of the adjacent side divided by the length of the hypotenuse, a/r. Since the hypotenuse of a right triangle is always longer than either of the other sides, both the sine and cosine are between -1 and 1, inclusive.

The sine is positive in the first and second quadrants and negative in the third and fourth. The triangle becomes degenerate if the point is on an axis but on the horizontal axis the b distance is zero, so sin(0) = 0. The cosine is positive in the first and fourth quadrants and negative in the other two. There are actually six important functions, including the *tangent*, but the others can be defined in terms of the sine and cosine.

The angle can be measured either in *degrees* or *radians*. A full circular angle is either 360 degrees or 2π radians. Visually the angle above is approximately 45 degrees or π/4 radians. The sine and cosine can be defined using either angle measurement, but you have to be clear in order to do calculations. Angles measured clockwise rather than counterclockwise are negative and angle measurement is periodic with period 360 degrees, so the above angle might also be 45 – 360 = -315 degrees, or 45 + 720 = 765 degrees, etc. The sine and cosine will be the same for all of these angle measurements, so sine and cosine are also periodic. Below is the graph of the sine function. The scaling of the horizontal axis is such that 1 is equivalent to 180 degrees or π radians.

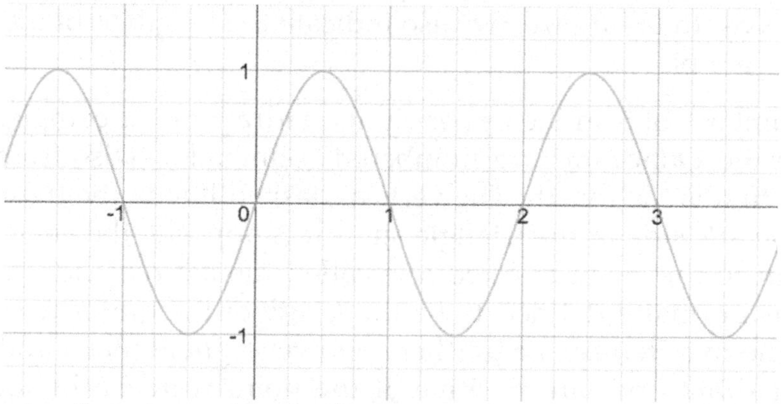

Below is the cosine function with similar scaling:

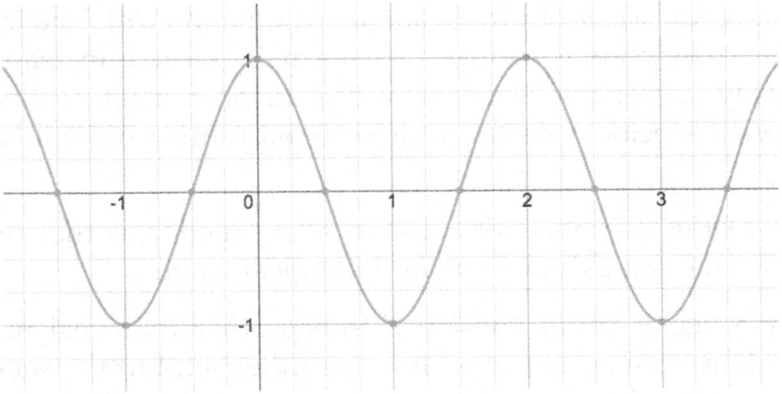

Also note that according to the definitions, a = r · cos(θ) and b = r · sin(θ). Therefore the value itself, z = a + bi = r(cos(θ) + i · sin(θ)). Thus the rules of complex numbers and the known facts about trigonometry can be used, each to aid the other. In particular, it is easy to show from the definitions given above that $\sin^2(\theta) + \cos^2(\theta) = 1$ and thus that the length of the vector cos(θ) + i · sin(θ) is also 1.

This also meshes with the theory of infinite series that we mentioned earlier, though you will need to wait for your Calculus course to explore that idea (Taylor's Series). Importantly, the infinite series give one way to compute estimates of the transcendental number π.

One of the lessons from all of this is that mathematics is very connected, though it has many fields with specialized techniques and different fundamental concerns, knowledge of one part can help explore and extend another. Indeed, the proof of Fermat's Last Theorem, mentioned earlier drew from many, seemingly unrelated, maths.

Foreshadowing – Matrix Theory

We can give an alternate definition of the Complex Numbers with a little background that also introduces another important field within mathematics.

Definition 11.4: The set $\{ k \mid k \in \mathfrak{N} \text{ and } k \leq n\}$ is called an *index set* of size *n*.

Definition 11.5: A *real vector* of dimension n is a function from the index set of size n to the reals. If the function is named v we normally write v_i instead of $v(i)$, similar to how we specify sequences. A vector of dimension n is also called an n–vector. We often express a specific n-vector as an ordered n-tuple of Real Numbers in the form $v = (v_1, v_2, \ldots, v_n)$.

A consequence of the definition of a vector as a function is that two vectors are equal if they have the same dimension and identical values.

More generally a vector could consist of things other than real numbers, as we shall see; a function from an index set to some other set. But the set of real vectors of a given length is, itself, an interesting mathematical object.

For example $(5, 3.2, 7, -\sqrt{2})$ is a Real vector of dimension 4; a 4–vector. If v_i are real numbers for i = 1..n then $v = (v_1, v_2, \ldots, v_n)$ is a real vector of dimension n. The v's are called the *components* of the vector.

Definition 11.6: The *sum* of two vectors v, and w of the same dimension, n, is the vector of dimension n whose components are the sums of the respective components of v and w: $v + w = (v_1 + w_1, \ldots, v_n + w_n)$.

We call such a sum (or other operation) *element–wise*.

Definition 11.7: The *dot product* of two real vectors v, and w of the same dimension, n, is the Real Number obtained by adding the products of the corresponding elements of v and w: $v \bullet w = v_1 \cdot w_1 + v_2 \cdot w_2 + \ldots + v_n \cdot w_n$. Note that the dot product is not a vector, but just a Real Number, if the vectors are real vectors.

A vector can be thought of as an ordered n–tuple or as a point in an n–dimensional space. The latter view has important geometric consequences, of course, though we won't pursue them here. Likewise the set of vectors of a given dimension (a *Vector Space*) are important in many fields of mathematics and in applications. Note that a Complex Number is a 2-vector of reals.

Definition 11.8: An n by m real *matrix* is an n–vector of m–vectors of reals. In other words take n distinct m–vectors as defined above and form a vector of them. Normally an n by m matrix is depicted as a rectangular array of reals. For example a 2 by 3 real matrix might be

$$\begin{bmatrix} 5 & 3 & 4.1 \\ -6 & 2 & -8 \end{bmatrix}$$

The components of a matrix z might be given as z_{ij} where i indicates that the component is on row i and the j indicates it is in column j. So in a 2 by 3 matrix we have components:

$$z = \begin{bmatrix} z_{11} & z_{12} & z_{13} \\ z_{21} & z_{22} & z_{23} \end{bmatrix}$$

Matrices have many applications throughout mathematics and physics. One use is to express geometric transformations, such as stretching, rotating, and translating of figures.

The two vectors making up the example matrix are (5, 3, 4.1) and (-6, 2, -8). However we can take an alternate view and consider the columns of the matrix to be vectors as well. In this view there are three 2–vectors, (5, –6), (3, 2), and (4.1, –8).

Definition 11.9: The sum of two n by m matrices is an n by m matrix whose components are the sums of corresponding elements of the two matrices.

Definition 11.10: The *scalar product* of a Real Number and a real valued matrix is obtained by multiplying each element of the matrix by the real number. The result is a matrix of the same shape as the original. A similar scalar product is defined for vectors.

Multiplication of matrices is more complex. The conditions under which we can even attempt it are strict.

Definition 11.11: The product of a k by n matrix, v, and an n by m matrix, w, is a k by m matrix, vw or $v \cdot w$ whose component on row i and column j is the dot product of the i'th row of v with the j'th column of w. If we call the product z then $z_{ij} = v_{i1} \cdot w_{1j} + v_{i2} \cdot w_{2j} + \ldots + v_{in} \cdot w_{nj}$. Note that i is unchanged for the v matrix and j is unchanged for w. In effect we match a row of the left matrix with a column of the right matrix and form the dot product to get a single component of the result.

This is a very strange definition when you first see it. The sum of vectors and matrices obeys many of the rules we have come to expect, such as the commutative and associative laws. A vector or matrix of all zeros acts like an additive identity. So vectors and matrix of a given size form a group. But matrix multiplication certainly isn't commutative, though it is associative and distributive over addition if the sizes match appropriately $w(x + y) = wx + wy$.

Thus, the product of a 2 by 3 matrix and a 3 by 4 matrix is a 2 by 4 matrix.

$$\begin{bmatrix} 1 & 2 & -1 \\ -3 & 4 & 5 \end{bmatrix} \begin{bmatrix} 1 & 2 & -1 & 1 \\ -3 & 4 & 5 & -1 \\ 0 & 2 & -1 & 2 \end{bmatrix} = \begin{bmatrix} -5 & 8 & 10 & -3 \\ -15 & 20 & 18 & 3 \end{bmatrix}$$

It is a bit hopeless to think of matrices in general as an algebraic structure, but the set of matrices of a given dimension do form an Abelian additive group. However, if we restrict ourselves to square matrices, in which the number of rows is the same as the number of columns, there is more hope since multiplication is at least *closed* in this set. There we find that matrices form a non–commutative ring. Multiplication has the "correct" properties, except that it isn't commutative. A matrix of all zeros is an additive identity, matrices have additive inverses, etc.

The (main) *diagonal* of a square matrix consists of those elements in which the two subscripts match; x_{11}, x_{22}, … x_{nn}. The square matrix with 1's on the main diagonal and zeros elsewhere acts like a multiplicative identity. But matrices in general don't have inverses. But most do; an interesting fact that you can explore.

That was a bit of a long build–up to a brief discussion of square 2 by 2 matrices with a certain restricted structure.

Suppose we only examine real matrices of the form $\begin{bmatrix} a & b \\ -b & a \end{bmatrix}$ where a and b are Real Numbers. The sum and products of such matrices also have the same form. Moreover, multiplication is commutative within this particular set.

It turns out that if either a or b (or both) is non–zero, then the matrix has an inverse. Remember that $I_2 = \begin{bmatrix} a & b \\ -b & a \end{bmatrix}$ is the multiplicative identity element and also has this form.

Consider $\begin{bmatrix} a & b \\ -b & a \end{bmatrix}$. Let $d = a^2 + b^2$. Assume that a or b (or both) is non-zero. Then $\begin{bmatrix} a/d & -b/d \\ b/d & a/d \end{bmatrix} \begin{bmatrix} a & b \\ -b & a \end{bmatrix} = I_2$.

The reason that these matrices are interesting at this point in our discussion is that $\begin{bmatrix} a & b \\ -b & a \end{bmatrix}$ behaves just exactly like $a + bi$, a complex number. The value i itself has form $\begin{bmatrix} 0 & 1 \\ -1 & 0 \end{bmatrix}$. And note the interesting fact that

$\begin{bmatrix} 0 & 1 \\ -1 & 0 \end{bmatrix} \begin{bmatrix} 0 & 1 \\ -1 & 0 \end{bmatrix} = \begin{bmatrix} -1 & 0 \\ 0 & -1 \end{bmatrix}$ which is the analogue of –1. The reals are embedded into these 2 by 2 matrices by associating a with $\begin{bmatrix} a & 0 \\ 0 & a \end{bmatrix}$. This is also just the scalar multiplication of a by I_2. So, we could have developed the complex numbers from the reals, by exploring such matrices, rather than the earlier approach.

I'll leave it there, but note a couple of facts. If you are a graphic designer, or a graphic game designer, you might want to take some graphic figure and rotate it, say counterclockwise through an angle of θ. If you treat the points of your graphic as column vectors of two dimensions, then multiplying that vector by the matrix $\begin{bmatrix} \cos(\theta) & -\sin(\theta) \\ \sin(\theta) & \cos(\theta) \end{bmatrix}$ will give you the coordinates of the new point. For example, for a right angle counterclockwise (90 degrees or π/2 radians) the matrix is $\begin{bmatrix} 0 & -1 \\ 1 & 0 \end{bmatrix}$. If the point is (3, 7), then the rotated point is the product $\begin{bmatrix} 0 & -1 \\ 1 & 0 \end{bmatrix} \begin{bmatrix} 3 \\ 7 \end{bmatrix} = \begin{bmatrix} -7 \\ 3 \end{bmatrix}$.

Such manipulations by matrix multiplication are very useful in geometry and even more so in a separate field of *projective geometry*(+). I leave it for your future explorations. Oh, but note that the rotation matrix is a complex number: $\cos(\theta) - i \cdot \sin(\theta)$.

There is an extension to the Complex Numbers, called Quaternions (William Rowan Hamilton, 1805–1865), that is useful in 3 dimensional geometry. Quaternions can be represented using specialized 4 by 4 matrices with real components or as specialized 2 by 2 matrices with complex components.

Summary So Far

We have seen only the briefest introduction to Complex Numbers, of course. The complete theory is very rich, however, and also very useful. But, by being algebraically complete, the Complex Numbers represents an end to our path that has been illuminated by those simple, polynomial, equations that have no solution in the simpler systems. But that is only one path through a small part of the vast landscape of mathematics that I've described earlier. There are many other possible paths that lead through, for example, Topology and Analysis, to yet richer systems.

Each time we expand we either open new vistas by making new things possible, or we contract, by making some things necessary that were only possible in the simpler world.

There are so many extensions and variations possible. A lifetime of exploration awaits.

12 Mathematics

What have we learned of mathematics itself here? What do mathematicians actually do? Would you like to do that?

What Do Mathematicians Do?

Mathematics is hard. Well, in some ways. Mathematics builds mind structures and it builds them on other mind structures. So to get in the game, you need to get a foundation on which you can build. Mathematics seems very methodical, but really isn't. It is chaotic at the edges where it really matters. Descartes seems to be the person who named imaginary numbers and thought of it as an insult. We wouldn't have modern electrical engineering, nor computers and smartphones without imaginary numbers, though. Someone thought they would be useful and started an exploration. At the time it had only theoretical interest. And it was interesting to only a few people. But the key step was that someone had the imagination to think it was worth exploring such a weird idea. But the same was true even for zero, back, way back.

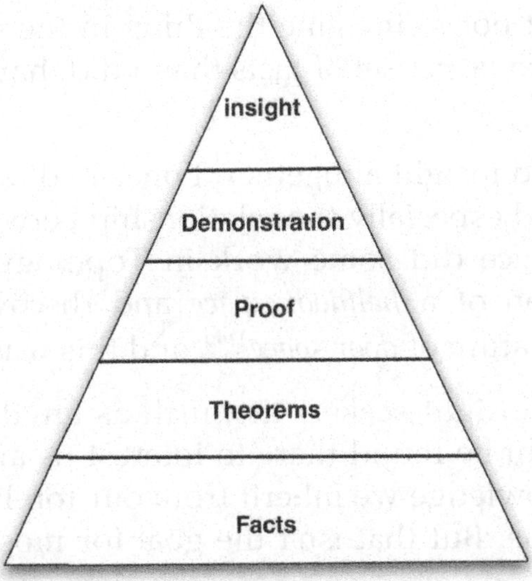

Consider this pyramid as representing math. Most people explore only the lowest levels. Most learn mathematics by working through this from the bottom up. In your childhood you learn a few facts. Then more and more in school. Only a few reach the top.

However, mathematicians actually work from the top of this structure down. The essence of *doing* math is to imagine what might possibly be true, or at least interesting to explore. The very tip of the pyramid is exploration. Often, even usually, their explorations come to nothing. In my doctoral student days I worked on three different problems looking for one suitable for a thesis. The first proved to be too easy. I could generate several theorems every day. The study was trivial and therefore uninteresting. The second problem was too hard. Neither my advisor nor I could get the slightest grasp. I abandoned it after a few weeks of frustration. The third was just right and led to my dissertation. It is an interesting (to me and a few others) extension to Calculus that is very much in the tradition of Newton and Leibniz, creators of Calculus, and all those who came between us. Mathematical interest moved in other directions for most people at about that time, so I may be one of the last in that particular tradition: Classical Real Analysis.

But, I cannot say that I had reached the top of the pyramid by then. I'd made a few excursions there, but hadn't really *got it* yet. All three of the problems I worked on were suggested by my advisor, Jan Mařik. He was the one with the insight. I was pretty good, though. Good enough to demonstrate some truths and prove important theorems. But not to imagine the thing in the first place. But of course, he had mastery of a much larger set of facts than I did, having been in the game for much longer.

That is not to say I lacked insight altogether. I once had deep understanding about real valued functions and especially the relationship between their large and small scale behavior. I also once did some work in Topology that I was proud of. I helped define the notion of a *halfdoor space* and discovered and proved some theorems. There is a literature of *door spaces*[+], and this was related.

So, mathematicians are insight seekers as much as anything. We peek into dark corners to see what might be found there to interest us and that might be used to extend the corpus of knowledge we inherit from our forebears. If what we do, and create, is useful, then fine. But that isn't the goal for most mathematicians. Many work in areas so specialized and arcane that it is a surprise if the results turn out to be useful, as I was surprised when informed, after many years, that my thesis actually had application.

But that doesn't explain, exactly, what mathematics *itself* is. Good definitions are very hard to find, and are generally wanting. But it is clear, at least, that without the human mind, mathematics would not exist. Mathematics is a purely mental construct, that can be useful, or just entertaining to the initiates. Mathematics is one of the classic studies and, along with philosophy, is a field of study meant to push the limits of understanding to their farthest limits.

Mathematics *could* have existed as much as two hundred thousand years ago, as that is when modern humans arose from our ancestors. But it probably didn't, though they had the same "brain power" as you and I. There are two reasons that it didn't arise until only a few thousand years ago. The people at that time had other pressing problems: Lions and baboons, and such. It was a very dangerous world in which the everyday problems of survival dominated everything. But the other reason is that there was, at the time, nothing of mathematics to build on. There were no facts to form the base of our pyramid. It took a long time, and a lot of leisure time, before maths could arise.

A Simple Example

Here I'll try to give a simple example of insight in action. When I was still in secondary school, the husband, Gene Richardson, of a cousin of mine introduced me to a way to compute square roots using only subtraction. I explored it quite a lot and wrote a paper on it, my first. I had too little background at the time to really nail the ideas down, but will here write a fictional account of how such a thing and its consequences might be developed. We have already seen some of the background above, in fact, but I'll start from the "beginning." As usual, it starts with a simple insight rooted in specific examples.

Someone starts out by noting that $1 + 3 + 5 = 9$ and $1 + 3 + 5 + 7 = 16$. Both of these are squares. They are the sum of the smallest positive odd numbers. Sixteen is the sum of the first four odd numbers but also the square of four. Is there a general pattern?

Yes, we have seen that, in general, n^2 is the sum of the first n odd numbers. The proof is by induction, though you need to know about induction in order to produce a proof. But you can approach the idea from a number of different angles. Someone is thinking and exploring. If you look at the sequence of consecutive perfect squares, 1, 4, 9, 16, 25, 36, etc., you might notice that the consecutive elements always differ by an odd number: 3, 5, 7, 9, 11, ... Now isn't that interesting. Of course, the two ideas above are really just different ways of saying the same thing, or nearly so. If we formalize it we get $n^2=\sum_{k=1}^{n}(2k-1)$. To get the next square, we add the next odd number. How odd. In particular, the first 10 odd numbers, 1 through 19, add to 100, and the sum of the first 100, 1 through 199, add to 10,000. But the same is true for 1000 and a million. Interesting.

Exercise: The sum of the first *n* even Natural Numbers in $n^2 + n$.

A bit later, perhaps this inquiring mind notices that $(mn)^2 = m^2n^2$. Putting these ideas together and adding up hundreds instead of units we see that 100 + 300 = 400 and 100 + 300 + 500 = 900. But the left side consists of odd multiples of 100 and the right side is a perfect square. In fact it is also 100 times another perfect square. So, this is $1 \cdot 10^2 + 3 \cdot 10^2 + 5 \cdot 10^2 = 9 \cdot 10^2$. But of course, it must and the sum of odd multiples of 10^4 gives a similar result for multiples of 10000: 10000 + 30000 + 50000 + 70000 = 160000 = 400^2.

Exercise: Can we say anything similar about the sum of multiples of odd powers of ten?

The sequence of consecutive odd numbers is very interesting. We have seen that prefixes of this sequence sum to perfect squares. But it is much more interesting. Suppose that you sum consecutive pairs of elements of this sequence: (1 + 3), (5 + 7), (9 + 11), etc. What you get is 4 times the corresponding element of the original sequence. Suppose you sum successive groups of three items instead: (1 + 3 + 5), (7 + 9 + 11), (13 + 15 + 17), etc. Hmmm, 9 times the corresponding element. Groups of four: (1 + 3 + 5 + 7), (9 + 11 + 13 + 15), etc. Now it is 16 times the corresponding original element. This is a clear pattern. You can, perhaps prove a theorem that adding the mth successive group of k odd numbers gives you k^2m. We will want this result later.

Exercise: What do you get if you group the terms ten at a time? One hundred at a time?

Here is another insight. If we look at the formula above, $n^2 = \sum_{k=1}^{n} (2k-1)$, we see a simple relationship between n^2 and the last number in the sum. The last value above is 2n – 1. So if we have a number like 625, which is 25^2 the last value must be $2 \cdot 25 - 1 = 49$. So, 625 is the sum of odd numbers from 1 to 49. Reversing this, if we have an odd number such as 87 then the sum of odd numbers from 1 through 87 must be the square of (87 + 1)/2 or 44 squared or 1936. That is, if $87 = 2 \cdot n - 1$, the last term in the sum, then n = 44.

But, suppose that we have added up hundreds instead of units. 100 + 300 + 500 + 700. The last multiple of 100 was 7, and (7 + 1)/2 = 4, but it is 4 tens that we want, so we have $40^2 = 1600$. But, turning it around again, the last odd number of the first forty is $2 \cdot 40 - 1 = 79$. Can we find a pattern here? Let's try another. Let's try to add six odd multiples of 100, ending with $11 \cdot 100$. We get 100 + 300 + 500 + 700 + 900 + 1100 = 3600 = 60^2 which is the sum of the first 60 odd numbers, the last of which is 119. Hmmm. That looks suspiciously easy 7 → 79, 11 → 119.

Since $100 \cdot n^2 = (10 \cdot n)^2$ we must also have that $100 \cdot \sum_{k=1}^{n} (2k-1) = \sum_{k=1}^{10n} (2k-1)$. The last value of the sum on the left is 2n-1 and on the right it is 2(10n) – 1 = 10(2n-1) + 9. So, the last odd number added if we just add units (2(10n) – 1) is ten times the last odd number of hundreds added, plus 9. So, it really is that easy. If I've just added a bunch of consecutive odd hundreds, say ending with 37 hundreds, that is the same as adding up odd numbers until we reach 379. 37 → 379.

Let's do a sanity check on the last sum. If I add up odd multiples of 100 until I reach 3700, I've added (37 + 1)/2 = 19 values so the sum is $19^2 \cdot 100 = 36100$. Doing it the other way, adding up odd numbers ending with 379, we add up (379 + 1)/2 = 190 values and $190^2 = 36100$.

Exercise: See what you can accomplish thinking about adding up odd multiples of 10000. And why doesn't it work for odd multiples of 1000?

Another way to look at the formula above, is that if I add consecutive odd hundreds, say *m* of them it is the same as adding $10 \cdot m$ individual odd numbers.

Here is another insight. Since subtraction is, in some sense, an inverse operation to addition, and we can use addition to compute squares, perhaps we can use subtraction to compute square roots. Interesting.

Now, let's see how this can apply to square roots. First, let's start with a simple case in which we know the answer already. Since $25^2 = 625$ we can compute the square root of 625 by repeatedly subtracting consecutive odd numbers until we reach 0.

```
 625
  -1
----
 624
  -3
----
 621
  -5
----
 616
etc. etc.
  96
 -47
----
  49
 -49
----
   0
```

There will be 25 values subtracted, of course, with $2 \cdot 25 - 1 = 49$ the last value. That was tedious, of course, but suppose we subtract hundreds instead:

```
 625
-100
----
 525
-300
----
 225
```

We can't subtract 500 from this if we want the result to be positive, of course, but note that the last hundred we subtracted was 3 (hundreds) so in effect we have odd number up to and including 39 from 625. Also note that we have made two subtractions. But since they were subtractions of hundreds it was just as if we had made 20 subtractions. The next odd number would be 41, so suppose we start subtracting from 225 starting with 41:

$$\begin{array}{r} 225 \\ -41 \\ \hline 184 \\ -43 \\ \hline 141 \\ -45 \\ \hline 96 \\ -47 \\ \hline 49 \\ -49 \\ \hline 0 \end{array}$$

Here we made five additional subtractions or the equivalent of 25 overall. But we only needed seven operations to do it instead of the 25 individual ones.

There is another way to reach the conclusion that we made 25 subtractions, actually. Note that the last number we subtracted to get 0 was 49. If this is $2n - 1$ for some n, then $n = 25$, of course. We have used this relationship before. So, the last number of hundreds was critical to discover the next number with which to continue, and the last number at the end will also give us the desired result. This is starting to get interesting.

We need to do a refinement if the number we start with has more digits. Notice that we started at the left and moved over after the subtractions of hundreds. We were working in the hundreds "place" and moved over to the "units" place: two decimal places.

Suppose we start with a much larger number, say 54756 (which just happens to be the perfect square of 234). Let's start at the decimal point and mark the number off in groups of two digits (hundreds, remember).

5 47 56

We start at the left and will, in effect be subtracting ten thousands, but won't write all the zeros.

```
 5  47  56
-1
--
 4
-3
--
 1
```

We can't make another subtraction, so we "bring down" the next group (47) to get 147. The last subtraction we made was 3 (ten thousands) or 39 (hundreds) so the next subtraction will be 41 (hundreds). Continuing:

```
 147
 -41
 ---
 106
 -43
 ---
  63
 -45
 ---
  18
```

Here we made three subtractions (or 23 overall) and the last one was 45 hundreds or 459 units. So the next odd number to be subtracted is 461 and if we bring down the next group (56) we start subtracting from 1856. Here we go:

```
 1856
 -461
 1395
 -463
  932
 -465
  467
 -467
    0
```

So, with four more subtractions we have made 234 altogether, so the square root is as advertised. Also, the last value subtracted, 467 = 2n - 1 gives the same result 234. QED

Actually there is another shortcut you can take to avoid a lot of subtractions. Note that if you are starting at 1856 and subtracting numbers starting with 461 there can only be three or four of them just with a simple estimation. Since they go up by twos, if there are actually four then the average of the four would have to be 464, midway between the second and third one. But four times 464 is, in fact 1856.

Let's try another. Suppose we want the square root of 10609.

```
 1 06 09
-1
 0
```

Now, the next subtrahend should be 21, but when we bring down the next group we have only 0 06. So we can't do the subtraction and need to bring down another group as well, so we want to subtract from 609. We haven't seen this situation before. In effect, since we couldn't subtract at all, we use 0 for the next digit of the square root, but what should we subtract next? Let's pause this and look again at how we shift. We noted above that when we subtract multiples of 100, ending with, say 1, the last subtraction is, in effect 19, so the next is 21. But suppose instead that we think of it as subtracting multiples of 10000. Since $10000 \cdot n^2 = (100 \cdot n)^2$ we get that $10000 \cdot \sum_{k=1}^{n} (2k-1) = \sum_{k=1}^{100n} (2k-1)$. Then the last odd integer subtracted in the formula on the right is 200n – 1. But this is the same as 100(2n – 1) + 99. So, if we last subtracted 5 ten thousands, then we have subtracted 599 odd integers. We need to extend the 5 with two nines. The next value to be subtracted would be 601 if we have "brought down" two groups in order to continue. The same pattern can be shown to be consistent. If you need to bring down 3 groups, extend with three nines, etc. There is a theorem hidden here, of course.

So, to continue with the example above we want to begin by subtracting 201= 199 + 2 from 609.

```
 609
-201
----
 408
-203
----
 205
-205
----
   0
```

There are three subtractions, so the square root of 10609 is 103, which you can easily check.

But, suppose we don't reach zero. Suppose we start with a number that isn't a perfect square. If we don't need to compute a rational estimate with some decimal places we can just stop when we can't go farther and we will have a rough (under) estimate, of course. Start with 630 and you will end up with twenty-five subtractions and a remaining 5 at the end. So the square root of 630 must be a bit more than 25. But suppose that isn't good enough. Suppose we do want to go farther.

Suppose that we want the square root of 5 to some number of decimal places to the right of the point. Note that we can write 5 as 5.00 00 00 00 00, using as many groups of zeros as we like. The process we have described above works just as well in this case. Except for where we place the decimal point in our result we can ignore that most of this work will be to the right of the decimal point. We will break the work up into parts as we go along.

```
 5 00
-1
 4
-3
 1 00
```

Thus, the first digit of our result is 2 and the decimal point will follow it. We already brought down the next group. The next subtrahend will be 41 = 39 + 2.

```
 100 00
 -41
  59
 -43
  16 00
```

So the second digit is 2; so 2.2... We continue, subtracting 441 = 439 + 2.

```
1600 00
-441
1159
-443
 716
-445
 271 00
```

The third digit is 3. So we have 2.23. The square of this is 4.9729, so we have an approximation. But let's continue with the next value to be subtracted which is 4461.

```
27100 00
 -4461
 22639
 -4463
 18176
 -4465
 13711
 -4467
  9244
 -4469
  4775
 -4471
   304 00
```

So, we have six subtractions and an estimate of 2.236 with 44721 as the next value to be subtracted. But this subtrahend is too large, so we need to bring down another group of zeros and recompute the subtrahend. After extending 4471 with two nines and adding two we get 447201, to be subtracted from 3040000. But we did no subtractions this time, so the next digit is 0 and our new estimate is 2.2360. The square of this is 4.999696 so we are getting very close. Let's do one more digit.

```
3040000 00
 -447201
 2592799
 -447203
 2145596
 -447205
 1698391
 -447207
 1251184
 -447209
  803975
 -447211
  356764 00
```

There were six subtractions, so our new estimate is 2.23606. We could leave it at that, noting that we have the exact decimals of the decimal expansion in the Real Numbers or we could note the following. If we bring down the next group we would be subtracting from 35676400 and would start by subtracting 4472121. But the latter number is approximately one seventh of the former, so we will have a large next digit. So, we could use 2.23607 as the rounded estimate to six places. This is a slight overestimate, however, so it depends on what you need to choose between the two values. But either is quite close to the square root of five. Below we can see the entire computation written out horizontally. Instead of "bringing down" the next group, we "bring up" the remainder from one computation to start the one to its right. The answer is above the computation.

2.	2	3	6	0	6
5	100	1600	27100	30400	3040000
-1	-41	-441	-4461	-44721	-447201
4	59	1159	22639		2592799
-3	-43	-443	-4463		-447203
1	16	716	18176		2145596
		-445	-4465		-447205
		271	13711		1698391
			-4467		-447207
			9244		1251184
			-4469		-447209
			4775		803975
			-4471		-447211
			304		356764

Some things are notable here. We never need more than nine subtractions in any column because of the adjustments made in the subtrahends. On the negative side, however, is that the values get larger and larger as you move to the right, adding to the work. Also, if we add one to the last subtrahend, it is twice the result, ignoring the decimal place.

Exercise: Earlier in this section we looked at what happens if you group the terms of the sequence of odd numbers. Explore the relationship between that idea and that of $\sum_{k=1}^{m\cdot n}(2k-1) = (m\cdot n)^2 = m^2\cdot n^2 = \sum_{k=1}^{m}(2k-1)\cdot\sum_{k=1}^{n}(2k-1)$.

So there you have it: Mathematics. Start with some facts. Add a little insight. Do some work, maybe a lot, to see if the insight tells you anything. Put it together with another insight or two and a few remembered facts. Demonstrate that your insights are sound, and, perhaps, prove a theorem or two. More work, some of it boring. Perhaps you come up with something useful, but, if it is *interesting* that is enough.

There was nothing earth shattering in the sample above. The insights were small. But some parts fit well together. So, satisfying. If it is useful enough it may reach the point in which school children are just taught the mechanics of it as a bunch of facts to be learned. Only later, if they decide to explore mathematics will they be led to their own insights, starting the cycle once again.

May all your insights be deep and most of your work not boring. Interesting.

13 Ideas for Studying Mathematics

Some people hate math. Some people think they can't learn it. A lot of this is poor teaching, perhaps by people who think that their students learn just like they, themselves, do. Part of it is inattention on the part of students. Missing an important idea can leave you mystified about ideas that depend on it. Some hints for studying Math follows. Most of them apply to the study of anything.

Learning in general is hard work. Math isn't fundamentally different. Learning has many levels. Seeing something once isn't learning. Remembering is helpful to learning, but not the whole story. To say you have really learned something you need to be able to apply the learning in different contexts. You need to make it useful, either for further learning, or in the real world. Your mind works in strange ways. Sometimes it works all by itself without conscious thought. But a good way to reinforce learning is to use repetition and to understand the difference between short term memory and long term. You should try to find ways to move important things from short term to long term memory.

Having important facts in your smartphone isn't the same thing as learning. Having Google at your fingertips can give you facts, but it isn't the same as learning. Again, you need to be able to *apply* what you know and the facts themselves are a poor substitute. The fact that most of mathematics is available online, and the wikipedia articles are very good, doesn't make *you* any smarter.

If you listen to a lecture or read a passage in a book or online, you haven't learned it. Some students have misconceptions about how necessary repetition is. Some students don't bother to take notes in class. Some think that, having heard or seen something, they will remember it; that they know it. Your mind isn't like a computer disk drive - put something in, it stays. Your mind works by physically rewiring itself. Connections need to be made between neurons. And they need to be reinforced.

Some students think that if they read a book and mark what seems to be important passages that it is enough. It isn't, and can be counterproductive. As you read you are not being sufficiently reflective about what you are reading. If you mark it on the third reading you will be better off, but then you will later need to return to your markup, not just leave it on the shelf. I've seen many markups in books where everything is marked. Not very helpful.

One important idea is that writing is more effective than reading. I mean writing with a pen or pencil on actual physical paper, not typing on a keyboard. It engages your mind more fully, which is essential to learning. If you take notes, and you must, do so on good quality paper that you can save and re-read. You may eventually want to review something you learned forty or fifty years ago. Trust me on that. And computerized notes are not really likely to be available to you in 40 years as the technology changes. Paper lasts. We can still read papyrus scrolls that are five thousand years old.

Some students think that if they take notes it is enough. It isn't, though it is a start. A much better scenario is to summarize your notes immediately after a lecture ends. What are the three most important ideas in a lecture? How can I be sure I remember them and, later, practice applying them? One good way to assure that you capture the good stuff is to carry a *Hipster PDA* and summarize every lecture on a small index card or two immediately after the lecture. What are the most important ideas in what I just heard. Then you can always carry a few cards, perhaps from different courses, so that you can review, and reinforce, whenever you have a moment.

If you use a trick like this to aid learning always carry a pen along with a small deck including a few blank cards. The other cards can be saved, re-arranged, or discarded as your needs require. When I was an undergraduate student the person who ended up at the top of our class always did this. I only developed the habit later. Both the writing out of the information and the constant review help you learn. It is named for the ability to carry it in a hip pocket without damage.

Another big idea about learning is that you are most likely to forget the thing you learned most recently. Capturing the key ideas immediately is important here, as is the review.

If you teach, you can give a Hipster PDA to every student. At the beginning of a daily class session you can ask several students for the big idea of the previous session. You can carry your own to aid in your lecture flow. The desk that I write this on is strewn with (too many) such cards. But the important ones tend to be on top.

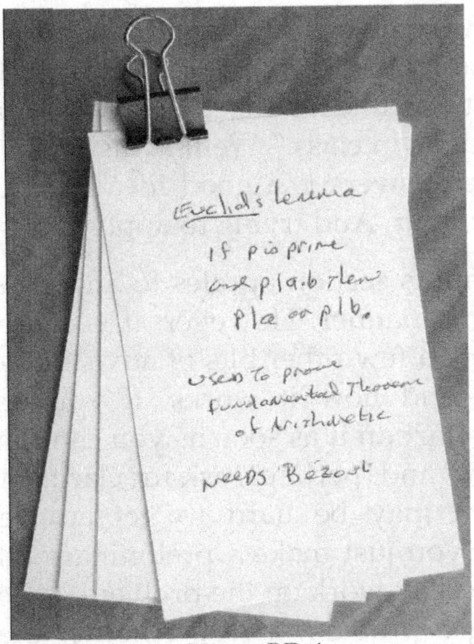

Hipster PDA

The other big idea about learning that I can share with you is that if you want to learn something, work hard at it. This is the same idea that applies if you want to be good at anything; swimming, cooking, writing, …

In mathematics, learning often means solving lots of problems. This is what first got me going in math and in school generally. It helped a lot that I liked solving geometry problems, but I worked on a vast number of them. I would seek to buy books of just problems and work them out for *fun*. It made a big change in my life. Later, trying to get good in Calculus, I graphed a huge number of rational functions (quotients of two polynomials) by hand, on paper. Calculus is a big help in this, of course, but I learned to recognize patterns. It helped me learn to look for places where functions in general misbehave, diverging to infinity, for example. It took a lot of time, but greatly increased my insight. With a graphing calculator, which didn't exist then, I might have been able to look at more such functions, but would likely have learned less. Doing it by hand was often tedious, but increased my ability; doing, not looking.

Asking questions is also very helpful. If you don't "get it" ask. My mother thought I was a pain for asking so many questions as a kid. My fellow students thought I was smarter than I was at the time for exactly the same reason. "Joe must really get it, he asks so many questions in class." Well, I was really just puzzled and wanted answers. Listening to the answers is a good idea too. And capturing the idea in your hipster. And reviewing it. And trying to apply it to some problem at hand.

Most of what I've said in this section applies to anyone studying mathematics or most other subjects in any manner. However, if you are studying in a school or university setting there are a few other bits of advice I'd offer. These are based on the nature of the brain and how it works. If you are given an assignment, especially a difficult one, start on it as soon as you can. This will give you a chance to find the difficulties early and, perhaps, ask for clarification as needed. If you are up against a deadline it may be hard to get advice. You needn't finish it immediately, but start. If you just make a preliminary attempt at the assignment, your brain will subconsciously work on the problem, even when you sleep. Insight often comes from giving a problem a rest. Another piece of advice deals with examinations. Staying up late, studying hard, cramming, the night before the big exam is actually counterproductive. It will leave you tired, confused, and possibly remembering the wrong things. You are much better off getting a good rest and arriving refreshed.

And the best advice I can give is to not be narrow in your studies. Study a lot of things, some of them deeply. This will both give you more options in selecting your path and will make you a better practitioner in whatever field you choose.

14 Further Reading

Appel and Haken, The Four Color Theorem.
https://en.wikipedia.org/wiki/Four_color_theorem

Bergin, A New Approach to an Old Problem, unpublished, 1961.

Bergin, A New Characterization of Cesàro–Perron Integrals Using Peano Derivates, TransAMS, Vol. 228, 1977

Cheng, The Joy of Abstraction, Cambridge University Press, 2023

Doxiadis and Papadimitriou: Logicomix: An epic search for truth, Bloomsbury USA, 2009

Gödel: The Incompleteness Theorems:
https://en.wikipedia.org/wiki/Gödel%27s_incompleteness_theorems

Landau, Foundations of Analysis, 1951

Index

Abacus....60
Abel, Niels Henrik....77
Abelian Group....77
Absolute value....83
Abstract Algebra....77
Additive identity....54
Additive inverse....83
Adjacent side....159
Aleph Nought....109
Aleph Null....109
Algebraic number....152
Algebraically Closed....157
Alternating harmonic series....121
Alternating series test....121
Antecedent....38
Antisymmetric....48
Appel, Kenneth....27
Applied Mathematics....78
Arabic Numerals....61
Associative Law of Addition....24
Average....106
Axiom Of Choice....41
Axiom of Induction....14
Axiom System....7

Bézout, Étienne....91
Bézout's Identity....90
Bijection....36
Binary....62
Bounded (sequence)....115
Bounded above....50, 114
Bounded below....115
Calculus....140
Cancellation Law of Multiplication....55
Canonical form....102
Canonical representation (of a Rational)....102
Cantor, Georg....138
Cardinality....33
Cardinality of R....138
Cartesian Product....34
Cauchy Process....124
Cauchy sequences....124
Cauchy, Baron Augustin–Louise....124
Choice problem....58
Chord....145
Classical Real Analysis....168
Closed (operation)....49
Closed interval....139
Closed set....139
Closure Axiom....11
Coefficient....151

Commutative Law of Addition....25
Complement....33
Complete (Real numbers are...)....134
Completeness....134
Complex conjugate....157
Composite....55
Conclusion....38
Conditional....37
Consequent....38
Constant polynomial....151
Constructive Mathematics....119
Constructive proof....119
Continuous (at a point)....142
Contrapositive....40
Converges to 0....115
Cosine....159
Countable....109
Countably infinite....109
Counting problem....58
Creativity Under Constraint....77
Curve matching....79
Decimal....62
Dedekind cuts....124
Dedekind, Richard....124
Degree (of a polynomial)....151
Degrees....160

Deleted neighborhood....141
Delta (as change)....143
Denominator....100
Dense....106
Derivative....143
Derivative (of a function)....145
Descartes, René....34, 167
Diagonal (of a square matrix)....164
Difference....81
Difference quotient....144
Dimension (of a vector)....162
Discrete....106
Discrete topology....140
Disjoint (equivalence classes)....68
Distance....106
Distance (between complex numbers)....157
Distinctness Theorem....16
Distinguishable....31
Distributive Law....54
Diverges....120
Dividend....86
Division....98
Divisor....86
Doing math....168
Door space....168
Dot product....162

Element of....31
Element-wise....162
Embed....43
Empty set....32
Epsilon-delta....125
Equality....11
Equivalence class....67
Equivalence classe....66
Equivalence relation....66
Euclid's Algorithm....90
Euclid's Lemma....92
Euler, Leonhard....89
Euler's Number....152
Even number....88
Existence (of numbers)....51
Existence proof....119
Exponent....151
Exponential Extrapolation....80
Extrapolation....79
Factor....54, 85, 88
Factorial....58
Fermat's Last Theorem....161
Field....111
For every....114
Four-color theorem....140
Function....35

Fundamental Theorem of Algebra....157
Fundamental Theorem of Arithmetic....93
General (statement)....39
Geometric series....121
Greater than....47
Group....76
Haken, Wolfgang....27
Halfdoor space....168
Hamilton, William Rowan....166
Harmonic series....121
Hexadecimal....62
Hipster PDA....182
Hypotenuse....159
Hypothesis....30, 38
If and only if....39
Image (of a relation)....36
Imaginary (number)....156
Imaginary part....156
Index set....161
Infinite....108
Infinite sequence....113
Infinite series....120
Infinity....41
Injection....36
Injection Axiom....12
Intersection....32

Inverse....34
Isomorphism....72
Kadison–Singer problem....94
Law of Cancellation....29
Law of cancellation for multiplication....85
Law of Subctitution....55
Law of Trichotomy....48
Learning....181
Least element....50
Least Squares Approximation....81
Left Distributive....54
Leibniz, Gottfried....168
Less than....47
Less–Equals....46
Limit (of a function)....141
Limit (of a sequence)....115
Limit (Sequence of Real Numbers)....134
Linear Extrapolation....79
Logic....37
Lowest terms....102
Magnitude (of a complex number)....157
Mařik, Jan....168
Matching....52
Mathematical Analysis....140
Matrix....162
Maximum element....50

Maximum element (of a sequence)....118
Mersenne, Marin....89
Midpoint....106
Minuend....81
Models....13
Monomial....151
Monotone (sequence)....115
Monotone decreasing....115
Monotone increasing....115
Monotone strictly increasing (or decreasing)....115
Monotonic....83
Multiple....55, 88
Multiplicative identity....54
Multiplicative inverse....102
Negative....74
Neighborhood....141
Newton, Isaac....168
Non-negative....74
Non-positive....74
Number Line....98
Number Theory....89
Numerator....100
Octal....62
Odd number....88
One-to-one....36
Onto function....35

Open interval....139
Open set....139
Opposite side....159
Ordered pair....34
Ordered pairs....65
Partial order....48
Partial sum....120
Partially ordered set (poset)....48
Peano, Giuseppe....7
Peano's Axioms....7
Pi....152
Point (in a topological space)....140
Point-slope....143
Polynomial....151
Positive....74
Postulate....7
Power set....33
Predecessor....17
Prefix....114
Prime....55
Product....52
Product (of matrices)....163
Projective geometry....165
Proof by Contradiction....17
Proper subset....31
Proposition....37

Pure Mathematics........78
Pythagorean Law........157
Pythagoreans........97
Quadratic Extrapolation........79
Quantifier........114
Quaternion........166
Quipu........60
Quotient........86
Radians........160
Real part........156
Real vector........162
Reciprocal (of a complex number)........158
Reflexive........46, 67
Reflexive Axiom........11
Relation........34
Remainder........86
Repeated roots........152
Richardson, Gene........169
Right Distributive........53
Ring........77, 95
Root (of a polynomial)........151
Rotation matrix........165
Scalar product........163
Second derivative........147
Semigroup........95
Sexagesimal........63

Sine....159
Singleton set....32
Sparse....106
Square matrix....164
Squeeze....116
Step function....142
Structure preserving....72
Subsequences....114
Subset....31
Substitution Law of Addition....24
Substitution law of multiplication....84
Substitution Law of Multiplication....56
Subtrahend....81
Successor....9
Sum (of matrices)....163
Sum (of vectors)....162
Surjection....35
Symmetric....67
Symmetric Axiom....11
Tail (of a sequence)....114
Tangent....160
Taylor's Series....161
There exists....114
Topological space....140
Topology....139
Total order....48

Total ordering....75
Transcendental number....152
Transfinite arithmetic....138
Transitive....67
Transitive Axiom....11
Trivial....128
Union....32
Unique Factorization Theorem....93
Unit element....54
Unknown....59
Upper bound....50
Variable....58
Vector Space....162
Zahl....69
Zeno of Elea....120
Zeno's Paradox....4, 120
Zero....4, 43

Made in United States
Orlando, FL
06 November 2026

90299265R00116